ECOLOGY AND EVOLUTION OF LEK MATING BEHAVIOR IN THE LONG-TAILED HERMIT HUMMINGBIRD

ORNITHOLOGICAL MONOGRAPHS

This series, published by the American Ornithologists' Union, has been established for major papers too long for inclusion in the Union's journal, *The Auk*. Publication has been made possible through the generosity of Mrs. Carll Tucker and the Marcia Brady Tucker Foundation, Inc.

Correspondence concerning manuscripts for publication in the series should be addressed to the Editor, Dr. Mercedes S. Foster, Department of Zoology, University of California, Berkeley, California 94720.

Copies of *Ornithological Monographs* may be ordered from the Assistant to the Treasurer of the AOU, Glen E. Woolfenden, Department of Biology, University of South Florida, Tampa, Florida 33620. (See price list on back and inside back cover.)

Ornithological Monographs, No. 27, viii + 78 pp.

Editor of AOU Monographs, Mercedes S. Foster

Editor of this number, John William Hardy

Special Associate Editors of this issue, Mercedes S. Foster, Department of Zoology, University of California, Berkeley, and Barbara K. Snow, The Old Forge, Wingrave, Aylesbury, Bucks, England

Authors, F. Gary Stiles, Escuela de Biología, Universidad de Costa Rica, Ciudad Universitaria, Costa Rica, and Larry L. Wolf, Department of Biology, Syracuse University, Syracuse, New York 13210

First received, 15 January 1976; accepted, 9 September 1976; final revision completed, 15 March 1977

Issued September 24, 1979

Price $8.50 prepaid ($7.50 to AOU members)

Library of Congress Catalogue Card Number 79-67267

Printed by the Allen Press, Inc., Lawrence, Kansas 66044

ECOLOGY AND EVOLUTION OF LEK MATING BEHAVIOR IN THE LONG-TAILED HERMIT HUMMINGBIRD

BY

F. GARY STILES
Escuela de Biología, Universidad de Costa Rica,
Ciudad Universitaria, Costa Rica

AND

LARRY L. WOLF
Department of Biology, Syracuse University,
Syracuse, New York 13210

ORNITHOLOGICAL MONOGRAPHS NO. 27

PUBLISHED BY
THE AMERICAN ORNITHOLOGISTS' UNION
WASHINGTON, D.C.
1979

TABLE OF CONTENTS

INTRODUCTION ... 1

Morphology and general biology of *Phaethornis superciliosus* ... 2

METHODS AND MATERIALS ... 3

Study Areas ... 3

Color Marking ... 3

Lek Observations ... 6

Foraging Data and Nectar Sampling ... 6

Collecting and Other Observations ... 6

LEKS, TERRITORIES, AND LEK BEHAVIOR ... 7

Sex of Lek Residents ... 7

Characteristics of Lek Sites ... 7

Song Perches and Territories ... 10

Vocalizations ... 12

Visual Displays ... 15

Matings, True and False ... 18

FORAGING PATTERNS ... 21

Food Plants ... 21

Interspecies Relations ... 25

Strategies of Flower Exploitation ... 28

Nectar vs. Insects ... 30

TEMPORAL PATTERNING OF LEK ACTIVITY ... 32

Breeding and Lekking Seasons ... 32

Daily Pattern of Lek Activity ... 35

LEK DYNAMICS ... 43

Seasonal Variation in Lek Composition ... 43

Survivorship and Recruitment of Lek Residents ... 49

Intermale Relations on the Lek ... 51

Inter- and Intralek Movements ... 56

DISCUSSION ... 58

Lek Evolution in *Phaethornis superciliosus* ... 58

Comparisons with Other Hermit Species ... 64

Evolution of Lek Behavior in Hummingbirds ... 69

ACKNOWLEDGMENTS ... 71

SUMMARY ... 71

LITERATURE CITED ... 74

LIST OF FIGURES

Figure 1. Map of La Selva property 4
2. Monthly variation in temperature and rainfall at Finca La Selva 5
3. History of spatial relations of three subleks in lek I from 1969 to 1972 8
4. Typical song perch 9
5. Relation between territory size and distance of clear visibility from central song perches 11
6. Location of four song types on lek II in 1971 and 1972 14
7. Visual displays of *P. superciliosus* given around song perches in lek territories 16
8. Number of species of food plants of *P. superciliosus* in good bloom in different months, according to use categories 24
9. Daily nectar production of eight plant species frequently visited by *P. superciliosus* 26
10. Cumulative use of food plants by *P. superciliosus* in relation to corolla morphology and use by other hummingbirds 27
11. Records of distances at which individual male *P. superciliosus* foraged from their territories 31
12. Estimates of the breeding season of *P. superciliosus* 33
13. Annual variation in calling activity among resident males on lek Ic 34
14. Daily variation in calling activity on leks II, Ia, and Ib for one day early and late in the 1970 lek season 36
15. Variation in daily calling activity of five males resident on lek Ic on 29 March, 1971 37
16. Variation in daily calling by male RY on lek Ic, end of 1971 lek season and beginning of 1972 season 38
17. Frequency of chase activity on areas observed during all-day watches at lek Ic 39
18. Seasonal variation in daily calling activity from one territory on lek Ic 41
19. Presence and territorial occupancy of male *P. superciliosus* on lek I, February 1971 to April 1973 42
20. Locations of lek Ia, "El Surá," and lek Ic, "Surprise," showing vegetation types, major landmarks, and distribution of major food plants 43
21. Locations of territories of males on leks Ia and Ic in March, May, June, July, 1971 44
22. Locations of territories of males on leks Ia and Ic in December 1971 and January–March, 1972 46
23. Locations of territories of males on leks Ia and Ic in April, May, and August 1972 47
24. Locations of territories of males on leks Ia and Ic in January and March, 1973 48
25. Proportions of *P. superciliosus* marked at one site that were recorded at other sites different distances away 56
26. Hypothesized relationship of dominance and/or fitness to lek position in leks of grouse and Ruff vs. leks of *P. superciliosus* 62

LIST OF TABLES

Table 1. Mensural characteristics of *Phaethornis superciliosus* 2

2. Numbers of territorial male *Phaethornis superciliosus* on each of four leks during February–March in five seasons 8

3. Observed temporal relationships of displays in 85 behavioral interactions in *Phaethornis superciliosus* lek territories 19

4. Morphology, period of good bloom, and hummingbird visitation of food plants of *Phaethornis superciliosus* 22

5. Feeding records of *Phaethornis superciliosus* at La Selva 23

6. Dominance relations of *Phaethornis superciliosus* at several food plants 28

7. Dominance relations of *Phaethornis superciliosus* and other hermits at different food plants 29

8. Observed flower visits by lek males and other individuals along a 450-m transect between lek I and the field station 32

9. Daily activity patterns of *Phaethornis superciliosus* in different habitats 32

10. *Phaethornis superciliosus* mist-netted at lek I, 1971–1973 40

11. Survivorship of resident male *Phaethornis superciliosus* within and between lekking seasons: lek I, 1971–1972 50

12. Age structure of *Phaethornis superciliosus* on lek I in 1971–1973 lekking seasons 50

13. Subsequent status of young male *Phaethornis superciliosus* first appearing on lek I in 1971 and 1972 52

14. Shifts in lek territories of resident male *Phaethornis superciliosus* in relation to previous status 53

15. Morphological and behavioral parameters of male *Phaethornis superciliosus* in relation to lek position 55

INTRODUCTION

In a lek mating system, several males gather in a localized area in the breeding season, where they display to other males and to females, the latter displays sometimes leading to matings. Male courtship assemblies occur in a variety of birds (e.g., Lack 1968), many frogs and toads (e.g., Emlen 1976), some mammals (Buechner 1961; Leuthold 1966; Bradbury, pers. comm.), and certain insects (Wynne-Edwards 1962; Alexander 1975). In anurans and some insects, the concentrations of males competing for mating privileges occur where females are predictably localized in space due to oviposition requirements. Although these aggregations may have a complex social structure (Duellman 1967; Parker 1970; Campanella and Wolf 1974; Emlen 1976), their evolutionary history is distinct from that of certain birds and mammals, in which the male aggregations are generated within the social system itself. The actual site of bird and mammal communal display grounds, or leks, is traditional within the population. Only a few of many potential sites are used over long periods. In most of the species, at least the initial stages of mating occur solely or principally on the leks. It generally is assumed that a number of males displaying together is more attractive or stimulating to females than would be the sum of an equal number of males displaying separately (Robel 1967); thus a male could potentially father more offspring by joining an assembly than by displaying or waiting for females solitarily.

In lek species an individual male, to maximize his chances of mating, must be available whenever a receptive female arrives on the lek. The evolution of leks thus probably is limited to species in which males can devote most of their nonmaintenance time and energy to advertising and defending areas and courting females as they arrive. This practically requires the partial or total emancipation of the male from the reproductive effort beyond the insemination of the female (D. Snow 1962, 1963). The strong selective pressure on males for effective aggressive and courtship characteristics, together with the emancipation of the male from reproduction, have produced exaggerated sexual dimorphism in many lek species (Sibley 1957; Wynne-Edwards 1962).

Hummingbirds and at least some species of grouse, shorebirds, manakins, cotingas, and birds-of-paradise meet the prerequisites for the evolution of leks (D. Snow 1963). Lek social systems apparently are not uncommon in hummingbirds, but seem particularly characteristic of the genus *Phaethornis*, the hermit hummingbirds. This is somewhat anomalous in that the species of this genus are rather dull colored, mostly without marked sexual dimorphism. Published accounts of *Phaethornis* deal with general descriptions of lek behavior (Nicholson 1931; Davis 1934, 1958; Skutch 1951, 1964a; Arp 1957; B. Snow 1974) and song types within and between leks (D. Snow 1968; Wiley 1971).

The pattern of resource exploitation of the population rarely has been integrated into any analysis of lek behavior (but see Kruijt et al. 1972; Campanella and Wolf 1974; Lill 1974, 1976; Pitelka et al. 1974). Hummingbirds are known to be highly dependent upon nectar resources for breeding (review in Stiles 1973), and competition for nectar in hummingbirds is well documented (e.g., Cody 1968; Wolf 1969, 1970; Stiles and Wolf 1970), yet these factors have not been considered in discussions of the evolution of lek behavior in hummingbirds. In this

TABLE 1

MENSURAL CHARACTERISTICS OF *PHAETHORNIS SUPERCILIOSUS*

Sample	Sex	N	x̄	SD	Range
Weight (g)					
Known lek residents	♂	146	6.08	0.25	5.5–6.7
Collected birds[1]	♂	32	6.13	0.25	5.6–6.7
	♀	29	5.87	0.23	5.4–6.3
Bill length—exposed culmen (mm)					
Known lek residents	♂	146	37.49	1.18	34.0–40.0
Collected birds	♂	32	37.68	1.01	35.5–39.5
	♀	29	36.10	1.23	33.0–38.5
Wing length—flattened wing (mm)					
Known lek residents	♂	143	59.96	1.46	55.0–64.0
Collected birds	♂	32	60.07	1.46	57.0–63.5
	♀	29	57.22	1.44	54.0–60.0
Tail length—central rectrices (mm)					
Known lek residents	♂	139	66.82	2.11	61.5–74.0
Collected birds	♂	31	66.61	2.38	62.0–71.5
	♀	27	66.54	2.26	62.0–70.5

[1] Sexed by dissection.

paper we describe in detail the lek behavior of the Long-tailed Hermit (*Phaethornis superciliosus*) in Costa Rica and show how its social system is intimately related to patterns of flower exploitation; these are in turn strongly affected by interspecific competition for nectar.

MORPHOLOGY AND GENERAL BIOLOGY OF *PHAETHORNIS SUPERCILIOSUS*

Phaethornis superciliosus is a medium-sized hummingbird (6 ± g) with a long, decurved bill (37 mm) and generally dull-colored plumage. The underparts are grayish brown anteriorly shading to buff on the abdomen. The dorsum varies from dull bronzy green to dull purplish bronze, the feathers of the lower back and rump having broad buffy edgings. The face is conspicuously patterned with a buffy postocular stripe, a whitish-buff malar stripe, and a buff stripe down the center of the throat; these contrast with the dark brown of the rest of the face and crown. The maxilla is black, the mandible mostly dull pinkish orange. The mouth lining is orange, and, together with the facial stripes, produces a striking pattern when the bill is open. The outer rectrices are grayish brown basally, and brownish black medially, with narrow buffy tips. The two central rectrices are greatly elongated and conspicuously white tipped.

Plumages of the sexes are identical; however, most individuals can be sexed by measurements (Table 1). The plumage of immatures differs from that of adults chiefly in having much more conspicuous buffy feather edgings on the back, crown, and tertials; no marked change in plumage color occurs with age. In addition, in the horny sheath of the maxilla immatures have corrugations that gradually disappear over a period of months (Ortiz-Crespo 1972), making it possible to determine the approximate age in months for many young birds. (See Stiles and Wolf 1974 for details of the aging procedure.) *P. superciliosus* lacks a postjuvenal molt; young birds first molt when about one year old. Adults have a single complete annual molt, the timing of which varies greatly among individuals but is quite constant from year to year in the same individual (Stiles and Wolf 1974).

The nesting of *P. superciliosus* has been studied in greatest detail by Skutch (1964a), and in general appears similar to that of most other hermits (cf. Ruschi 1950; Skutch 1951; D. Snow and B. Snow 1973; B. Snow 1974). The female builds the nest, incubates, and rears the young by herself. Presumed males appear very rarely at nest sites and apparently play no constructive role in nesting (cf. B. Snow 1974). Nests are placed on the underside of the tips of palm or cyclanth leaves, or strips of *Heliconia* leaves; the nest has been described by Skutch (1964a). Nests are found most often along trails or streams, usually well away from leks (Skutch 1964a; B. Snow 1974; present study). Nesting success probably is low, perhaps 15–25%, as is the case in most other hermits studied to date (Skutch 1966; B. Snow 1974).

Phaethornis superciliosus inhabits the humid tropical and lower subtropical zones from southern Mexico through eastern Bolivia and Amazonian Brazil (Meyer de Schauensee 1966). Scattered reports in the literature and our own observations indicate that it is a lek species throughout this extensive range (Nicholson 1931; B. Snow 1973b; Wolf and Stiles, pers. obs.).

METHODS AND MATERIALS

Study Areas

Most of this study was done at Finca La Selva, in the Sarapiquí lowlands of northeastern Costa Rica (10° 26′ N, 84° 01′ W; elevations 50–150 m). About 75% of the roughly 700 ha of the La Selva property is primary tropical wet forest, with sizeable areas of old cacao plantations and second growth on the northern and eastern edges of the property (Fig. 1). Much of the land north, east, and west of La Selva has been cleared within the last 15 years, but the forest to the south is still continuous up to the volcanoes of the Cordillera Central. The forest at La Selva is described in detail by Holdridge et al. (1971); the second-growth and light-gap vegetations are discussed by Stiles (1975) with particular reference to *Heliconia*. The avifauna of La Selva has been studied in general terms by Slud (1960).

Rainfall at La Selva averaged 4,150 mm per year between 1960 and 1969. A rather poorly defined dry season extends from about late January through April in most years; months of heaviest rainfall usually are June–July and/or November–December. Temperature shows little seasonal variation with monthly averages of 22.8–24.9°C and a maximum range of 14°–37°C (Fig. 2).

A few supplementary observations were made near Rincón, Peninsula de Osa, in southwestern Costa Rica. This is another lowland wet forest area of similar climate and vegetation (described by Holdridge et al. 1971).

Color Marking

Nearly all of the observations reported here were made on color-marked individuals. Birds were captured with mist nets at leks or feeding areas. The

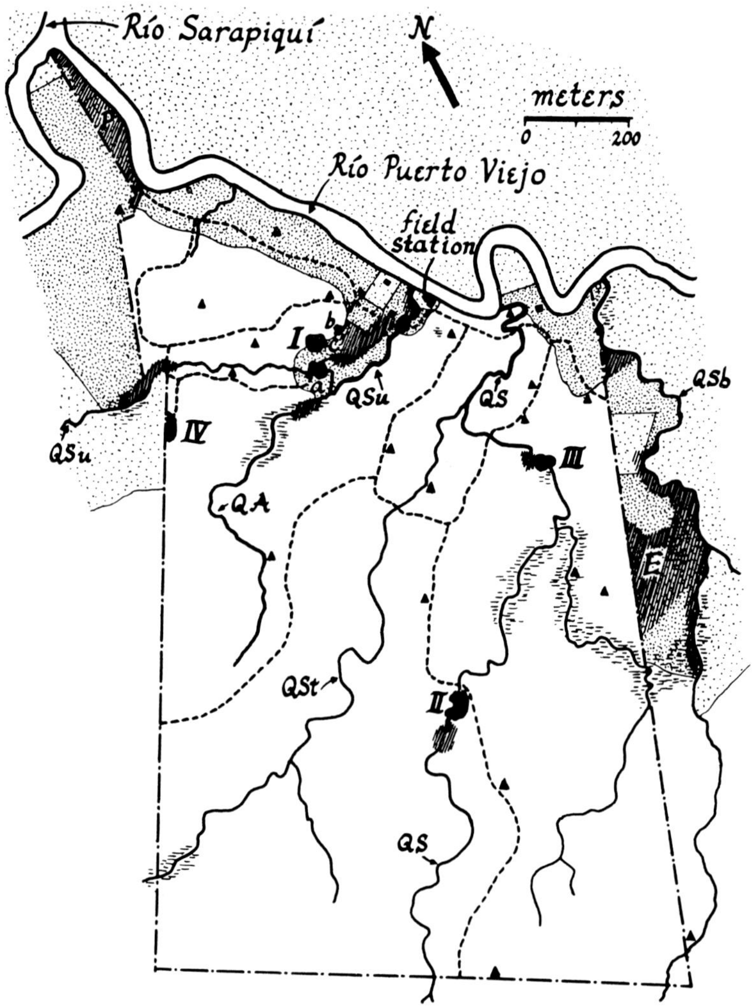

Fig. 1. Map of the La Selva property. Lek sites are indicated by Roman numerals (I–IV); diagonally hatched areas are feeding sites; nest locations are given by triangles; swampy areas are given by horizontal dashes. Abbreviations are, streams: QSu = Quebrada El Surá; QS = Quebrada El Salto; QSt = Quebrada El Saltito; QSb = Quebrada Sábalo; QA = Quebrada Arboleda; major feeding areas: P = Point; E = East Boundary. Cacao and second growth are shown by heavy stipple; pastures are light stipple.

back feathers were marked with spots of model airplane paint, and a plastic tag with stripes of colored plastic tape was affixed to the leg. Each individual was given a unique color combination. Further details of the marking method are given by Stiles and Wolf (1973). The paint was lost at each annual molt, and tags fell or broke off from time to time, so we attempted to recapture individuals one or more times each year to replace their markings. At two leks we netted at approximately monthly intervals during 1971–73 to follow in detail the progress of the annual molt in lek males (Stiles and Wolf 1974). Over the course of this

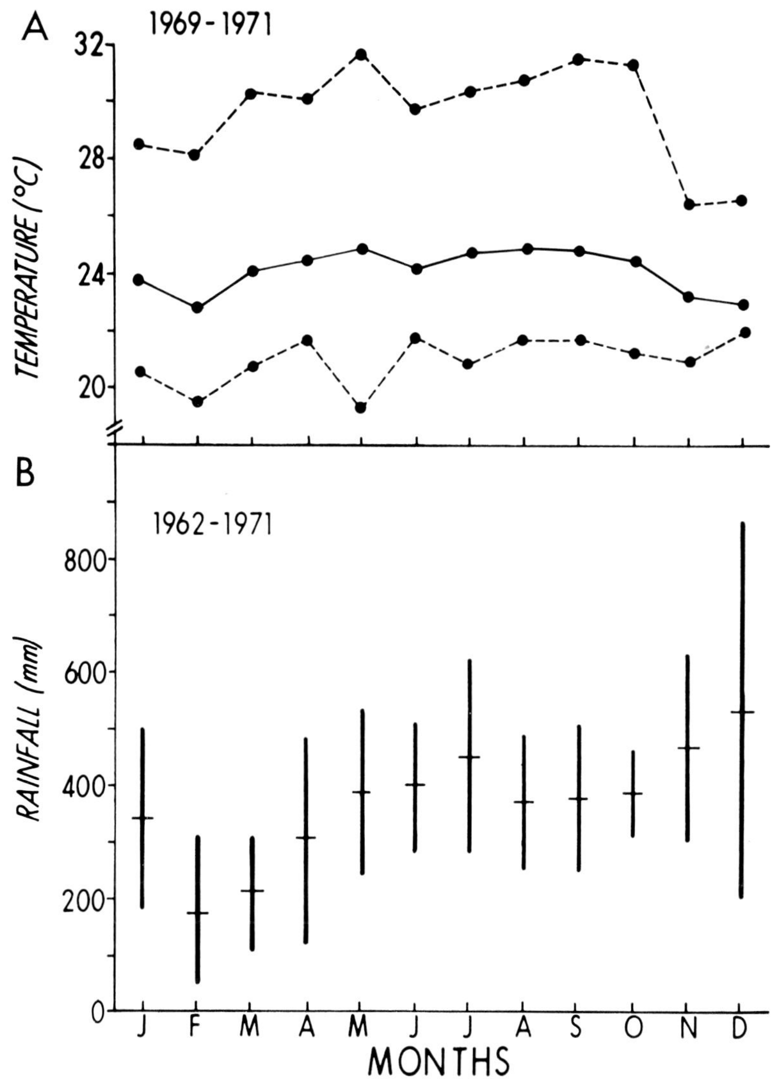

Fig. 2. Monthly variation in temperature and rainfall at Finca La Selva. A. Monthly mean maximum, mean, and mean minimum temperatures, reading from top to bottom. B. Monthly mean rainfall, plus and minus one standard deviation.

study, we marked over 500 individuals. From motion pictures and observations we detected no adverse behavioral effects of our marking system. Displays of marked and unmarked birds were indistinguishable, and the presence of markings on one individual did not appear to affect the behavior of individuals interacting with it. In a few individuals, mostly near the beginning of the study, perching ability and possibly survival were impaired by an improperly attached tag. Data from these individuals were excluded from the analyses.

Lek Observations

Observations on behavior of lek birds were made with 7× and 10× binoculars at 3–20 m ranges by an observer usually seated and partly concealed by the undergrowth. We found that after a short period of habituation the birds ignored the observer, provided he remained nearly motionless. Detailed reports of behavior were made on the spot, often recorded by a second person while the first watched. We also made 16 mm motion pictures of behavioral interactions that provided documentation of our field descriptions and were used to draw Figure 7.

We made periodic censuses of the leks during this study, attempting to locate all frequently used song perches of each male and to map the extent of each territory. We also made all-day singing records at approximately monthly intervals for a year at one lek and less frequently at two other leks. For these data the observer recorded singing or absence of singing by each individual of a group of 5–7 adjacent residents (or for the entire lek when this comprised 5 or fewer birds), during each minute of the day. All chases originating in or passing through the sector of the lek also were recorded. These data provide a quantitative index of lek activity through the day for different seasons.

Foraging Data and Nectar Sampling

We recorded foraging at flowers by *Phaethornis superciliosus* in our regular census walks through various forest and second-growth habitats and whenever we walked certain trails passing through several habitats. Each visit of a bird to the flowers of a plant or clone was treated as a separate feeding record. These data give a representative picture of flower use through the year by the *P. superciliosus* population at La Selva. We carried out regular netting and observation programs at certain major foraging areas during periods of blooming of important food plants, to document use of these sites by lek residents.

We made monthly censuses of the flowering of the food plants of *P. superciliosus* from 1971 through 1974. The average period of "good bloom" (= 50% or better of maximum observed flowering during this period) was used to assess overall flower availability to *P. superciliosus* (for further details see Stiles 1977, 1978a).

We sampled nectar of several important food plants of *P. superciliosus* with fine capillary tubes at regular intervals through the day. Flowers were bagged to prevent removal of nectar by birds or insects. Data on nectar production reported here were from days with clear mornings, to minimize variations in production rates due to weather.

Collecting and Other Observations

During 1971 and 1972, we made regular trips to a locality about 1.5 km from La Selva across the Rio Puerto Viejo. We collected 63 specimens of *P. superciliosus* to document gonad cycles and to obtain measurements of individuals of known sex. Measurements of males collected from this population were virtually identical to those of known lek males at La Selva; hence the measurements so obtained should apply equally well to both sexes of the La Selva population. In addition, nine actively calling lek residents were collected from nine different territories on a lek of *P. superciliosus* near Rincón de Osa at about 0800 on 8 August 1969. These birds were collected within a 10-min period to document the sex of lek residents.

Specimens collected during this study are deposited in the American Museum of Natural History, the Western Foundation of Vertebrate Zoology, and the Universidad de Costa Rica.

LEKS, TERRITORIES, AND LEK BEHAVIOR

Sex of Lek Residents

In the absence of discernible plumage dimorphism in *P. superciliosus*, sex of live birds was determined by external measurement. Males averaged heavier and had longer bills and wings than did females (Table 1). Tail length was not sexually dimorphic. Although there is overlap in all measurements, a combination of two or three enabled us to determine the sex of most individuals with a high degree of certainty. Measurements of 146 marked individuals known to be lek residents agree precisely with measurements of known males collected at a nearby locality (Table 1). Eight known lek residents at La Selva sexed by dissection were all males. Of the 146 lek residents measured (Table 1), 125 (86%) were unequivocally males on the basis of measurements. Only 5 (3%) would probably have been called females as all of their measurements fell near the low end of the observed range for males; one of these was sexed as a male by dissection. Furthermore, nine lek residents collected on a lek near Rincón de Osa were all found to be males upon dissection. Thus the evidence overwhelmingly indicates that in *P. superciliosus* all lek territory holders are males (see also Nicholson 1931). This also is the case in other species of *Phaethornis* (Arp 1957; B. Snow 1974, pers. obs.).

Characteristics of Lek Sites

We found four leks of *P. superciliosus* during this study. They were scattered over the La Selva property, the average distance between them being about 1 km (Fig. 1). We feel sure this is a complete inventory of the leks on the northern two-thirds of the property, which we searched intensively in February, March, and April 1972. However, we strongly suspect that a fifth lek exists near the southwestern corner of the property. The density of leks in the La Selva forest is thus about one per km^2, a surprisingly low figure because *P. superciliosus* is one of the commonest hummingbirds of La Selva.

When discovered in June 1969, lek I was comprised of two semi-isolated "sub-leks" about 150 m apart. The intervening area, which contained cacao plantations and a 1.5-ha banana patch that had been cleared in March 1968 (R. Chavarría, pers. comm.), was not suitable lek habitat in 1969. Prior to the clearing of the banana patch in 1968, the two subleks probably were much closer together, if not contiguous (Fig. 3). The larger "El Surá" sublek (Ia) was located along the Quebrada El Surá; in June 1969 it was comprised of 8 birds, and gradually declined thereafter (Table 2). The smaller "Banana Patch" sector (lek Ib) contained three birds during most of 1969 and 1970, with two birds being added along the edge of the banana patch toward lek Ia by July 1970. By February 1971

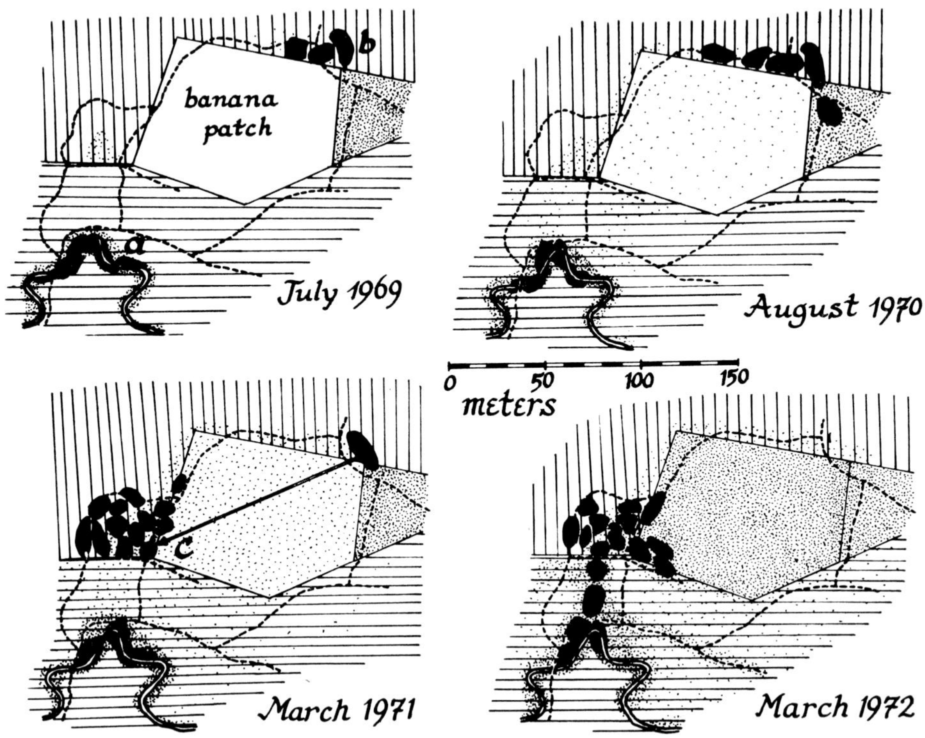

Fig. 3. History of spatial relations of three subleks in lek I from 1969 to 1972. From two rather isolated sections in 1969 the lek has gradually changed to one continuous group of birds as intervening areas undergo secondary plant succession. Vertical hatching = forest; horizontal hatching = cacao; stipple = dense thickets, the heaviness of the stipple corresponding to the density of the vegetation.

the third or "Surprise" sector (Ic) had been established about midway between the first two subleks. Lek Ib concurrently declined to a single bird, which shifted to lek Ic, leaving Ib empty. Sectors Ia and Ic gradually merged into a single large lek by mid-1972 (Fig. 3) as undergrowth grew up in the intervening area.

Leks II (discovered 1969) and III (discovered 1972) both lie along the Que-

TABLE 2

NUMBERS OF TERRITORIAL MALE *PHAETHORNIS SUPERCILIOSUS* ON EACH OF FOUR LEKS DURING FEBRUARY–MARCH IN FIVE SEASONS

		1969	1970	1971	1972	1973
Lek number	Ia	9	8	3	3	–
	Ib	3	5	(1)	–	–
	Ic	–	–	13	15	18
Total lek	I	12	13	16	18	18
	II	14[1]	19	23	19	22
	III	NT[2]	NT	NT	8[1]	16
	IV	NT	NT	NT	6	5

[1] Incomplete census.
[2] NT = not taken.

Fig. 4. A typical song perch (male GR on lek Ic). Note the open space around the song perch itself which is a slender, fairly horizontal twig or vine. The back paint and the tip of the tag are also visible on this marked male. See Figure 21 for the locations of this male in relation to the rest of the lek.

brada El Salto, about 1 km apart (Fig. 1). The former contains 20–25 birds, the latter about 15 birds (Table 2). Lek IV is small (Table 2) and is located on the west boundary of La Selva approximately 600 m from lek I (Fig. 1).

Three of the four leks are located at least partly along streams, which represent the largest and longest-lived gaps in the forest canopy at La Selva (Stiles 1975). This is important for three reasons: (1) Where light reaches the forest floor dense thickets can grow. (2) Streambanks provide habitat for several important food plants of *P. superciliosus*. (3) The streams are convenient flight paths for residents and visitors. All four leks are located in dense thickets (Fig. 4), along a man-made edge (lek IV), a stream (leks II and III), or a combination of these (lek I). The rise of lek Ic probably resulted from invasion of the banana patch by second growth and the growth of dense thickets in the adjacent forest and cacao (Fig. 3). The thickets in which lek IV is located almost certainly date from the clearing of the adjoining tract of land and the resultant penetration of the sun inside the forest along the edge. The most likely site for this lek prior to 1971 was just west of La Selva, possibly along the Quebrada El Surá.

The two leks in undisturbed habitat, II and III, are situated near open swamps along the Quebrada El Salto. *Heliconia pogonantha*, the most important food plant of *P. superciliosus*, is common in such swamps; this suggests that another important requisite of a lek site is proximity of good foraging areas. The situation of lek I appears to support this inference, although habitat disturbance is a complicating factor. A large open swamp with much *Heliconia* is some 350–400 m from the lek up the Quebrada Arbolera; the banks of both this creek and the Quebrada El Surá support good stands of several species of *Heliconia* (Stiles 1975). Moreover, lek I has increased in size in recent years as thick second growth containing much *H. pogonantha* has grown up in the banana patch. The location of lek IV is probably atypical. However, *H. pogonantha* was common along the Quebrada El Surá west of La Selva in 1968, and this area, now cleared, may then have been an important foraging area for the birds of lek IV.

Song Perches and Territories

The foci of aggressive and courtship activities on a lek are the song perches of the resident males (Fig. 4). A male on the lek spent most of his time on his song perches, singing, displaying, preening, or sitting quietly. These song perches varied from 1 to 7 m above ground, but most were 1.5–2 m. A male often used a high, relatively exposed perch during the dim light of dawn and then moved to lower, more shaded perches as the sun rose. The perches were typically bare, slender twigs or vines of 1 cm diameter or less, horizontal or nearly so. Usually the perches were located in a small clear space 0.5–1 m in diameter, amidst thick vegetation, allowing the resident a good view of his immediate vicinity and providing room for the various displays that involve flying near the perch. Males often used the same perches for weeks or months. Moreover, they would nearly always sing from precisely the same spot on the twigs, and usually facing the same direction. Other males (and females?) hovered around these perches as they flew through the lek, even when the resident was absent. It is probable that most or all of the regular song perches of all other residents are known to each male on a lek or lek sector.

The territory of a lek male typically consisted of two to five song perches and the area around and between them. There was no fixed boundary; the airspace of a territory was defended much less consistently than the song perches themselves. Some expansion and contraction of territories occurred on a day-to-day basis, reflecting the dynamics of a male's interactions with his neighbors. These changes in areas always involved the taking up or abandoning of song perches on the periphery of the territory; males did not defend areas that contained no song perches. Other males sometimes were permitted to fly through the territory by the resident, but never to sing or to use a song perch unless the resident was absent. At times, a neighboring resident chased an intruder from a song perch in the absence of the resident of the territory in question.

Territory size varied from 20 m^2 to almost 500 m^2 and was closely correlated with the openness of the vegetation as perceived from the song perches (Fig. 5). The smallest territories were in the very dense thickets adjoining breaks in the forest canopy such as streams, treefalls, open swamps, and edges. Territories tended to be closer together in such situations, leading to a considerably higher

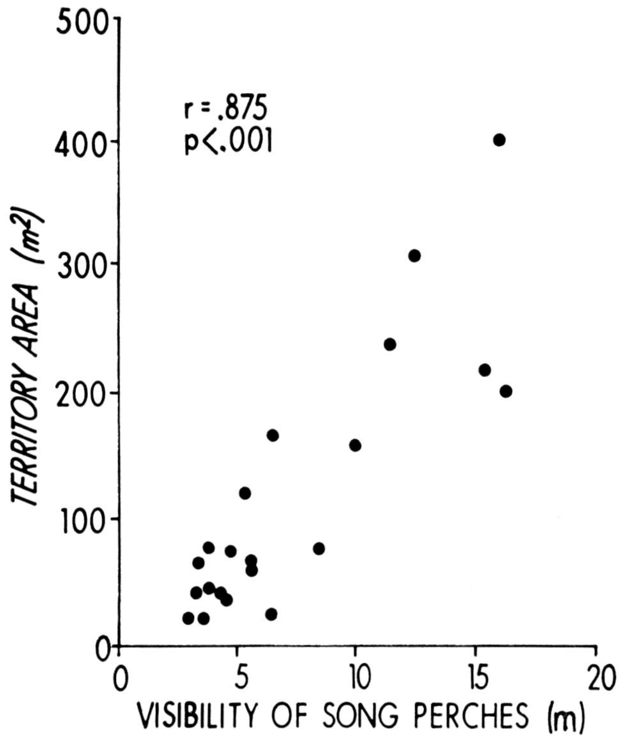

Fig. 5. Relation between territory size and distance of clear visibility from central song perches. Visibility distance is the average maximum distance at which the paint and tag markings of the bird on its perch were visible to an observer from several angles.

density of individuals. The largest territories were in relatively open forest understory or cacao, usually towards lek peripheries. Openness of the vegetation influences the ability of a resident to detect intruders and defend his song perches, and may set the minimum spacing between territories. Two adjacent residents on their song perches were rarely in visual contact. The few exceptions usually involved the temporary territorial shifts mentioned above and were accompanied by increases in aggressive interactions, notably chases. Constant visual contact among several residents may increase aggressive activity on the lek. This could have at least two unfavorable consequences: (1) a high degree of interference with mating attempts, and (2) an extra energy drain on the males.

Lek territories of *P. superciliosus* seldom contained flowers that were used as nectar sources by territory owners. Nearly all individuals we observed did all of their feeding away from the lek. A few males had suitable flowers on their territories that they visited but did not defend consistently against other hermits. In at least two cases, flowers on a male's territory were used mostly by hummingbirds of other species. One lek male abandoned part of his territory as a clump of *Heliconia imbricata* came into bloom there and was occupied by a territorial male *Thalurania furcata*. The *Thalurania* chased the *Phaethornis* whenever the latter attempted to use a song perch located 40 cm from an *H. imbricata* inflorescence.

Phaethornis superciliosus also eat small insects and spiders, obtained by foliage gleaning. Some males gleaned occasionally in their territories, but never frequently or regularly. It seems evident that territories were not being defended for their food value, but were solely display grounds that the males left to obtain virtually all their food. In all essentials the functional characteristics of *P. superciliosus* territories agree with those of the 'classical' lek territories reported for a variety of birds and other animals.

Vocalizations

We recorded a variety of vocalizations in *P. superciliosus*; most were heard chiefly or only on the lek. We made no attempt to study vocalizations of juveniles or females at the nest. Hence the following catalogue of vocalizations of *P. superciliosus* probably is not complete.

Flight call.—The flight call is a squeaky, whistled *sweet* or *tseet*. This was the vocalization most commonly heard away from the lek. It was given by a flying bird in a variety of circumstances: upon arriving at flowers, upon sighting a conspecific, upon seeing the observer or a predator, or sometimes with no obvious external stimulus. The flight call appears to announce a bird's presence, and it probably is analogous to the chip notes of certain North American hummingbirds in being a general contact and spacing note (Stiles 1971b).

Aggressive squeals and chatter.—These consist of one to four high, thin squeals, ending in a rapid ascending or descending chatter. Each successive squeal is lower in pitch than the preceding one; sometimes the chatter is omitted, or it may be followed by one or two ascending squeals. This call usually was given by a male perched either on his territory or by a flower clump, upon the approach of another hermit. We also once heard a female at her nest give this or a very similar call when another hummingbird flew by. The squeal and chatter appear to serve as a warning or threat, signifying the readiness of the calling bird to attack the intruder. Should the latter not leave, a chase would often follow. Sometimes squeals and/or chatters were given during the chase, apparently by the attacking bird (presumably the resident during a chase on the lek).

Chase notes.—These are a variety of staccato *pip* or *chip* notes given during a chase. When the chase occurred on the lek and involved two *P. superciliosus*, it was impossible to tell whether chase notes were being given by the chaser, the bird chased, or both. However, when a *P. superciliosus* was chased by another hummingbird (e.g., from flowers defended by the latter), it often gave chase notes, suggesting that in lek encounters the bird chased usually gives these notes. Lone birds flying rapidly through the lek, not involved in any chase, gave similar notes. The function of the chase notes remains unclear at present, and they do not seem to be strictly confined to chases.

The preceding vocalizations were heard both in and away from leks, and probably were given by both sexes. Those that follow were heard mainly on or near leks, and appeared to be given chiefly or only by males. The song was the most characteristic lek vocalization, but several others, including squeals, chatters, and chase notes, were given in aggressive encounters.

Song.—The song of this species consists of a single phrase, repeated monotonously for periods of up to 30 min. Singing occupied most of a male's time on territory and was his major means of announcing his presence. A male often spent up to 50% of the daylight hours singing during the early part of the lekking season.

A male singing on his territory in the absence of conspecifics usually gave 60–70 songs per min. When an intruder approached, this increased to 90–110 songs per min, and the songs became louder and shriller. Neighboring residents often countersang, alternating their songs with nearly perfect regularity. The singing rate of each such male usually dropped by about half, so that the number of songs per min by the two males combined was only slightly greater than the number given by a male singing alone.

Nonresident males attempting to invade or gain control of a lek territory also sang, but at least initially their songs were soft and very irregular in tone and rhythm, particularly when the invader was a young bird. If the nonresident went unchallenged for any length of time, his songs became louder, more uniform, and regular. Thus the rate, volume, uniformity, and rhythm of the songs provided a fairly good index of the stimulus input, aggressive state, and territorial status of the singing male.

We found a number of distinct song types in the *P. superciliosus* population of La Selva. Some idea of the distinctness of these dialects can be gained from some of our transliterations: *chink*, *churk*, *kaching*, *chrrik*, *shree*, *beeyurr*, and *kisink*. The number of song types represented on a lek was related to lek size. Lek II, with 20 or more residents in most years (Table 1), had 4 distinct song types. Lek Ia (5–8 residents) and Ib (3) had 1 song type each, while Ic, formed in part from the amalgamation of these 2 leks and containing approximately 15 birds, had 2. In 1973 lek III had 16 birds and 3 song types, lek IV 5 and 1, respectively. Within a lek the individuals singing a particular song type usually constituted a geographically defined subunit of the lek. These subunits maintain their integrity from year to year in spite of a considerable turnover of adult males (Fig. 6). On large leks with several subunits, members of one subunit sometimes could not hear members of another, although intervening subunits normally connected the two.

Similar groupings of song types occur within leks of *P. guy* and *P. longuemareus* (D. Snow 1968; Wiley 1971; B. Snow 1974). Further details regarding song dialects in *P. superciliosus* are presented elsewhere (Wolf and Stiles MS). For the present, suffice it to say that individuals appear to learn their songs in their first year and sing the same song thereafter.

Chip-churr.—This call consists of one or several staccato notes, often on different pitches, followed by a nasal, rolling *churr* or *chirr*. Sometimes the *chip* notes were omitted, and several *churrs* were given consecutively. The *chip* notes are quite similar to chase notes; however, the *churr* was never heard in a chase. The *chip-churr* usually was given by a male on his song perch in a variety of situations, e.g., when a conspecific or a chase came through the territory, when a potentially dangerous animal (squirrel, weasel, human being, etc.) was near, or preceding flight in the apparent absence of other hummingbirds or animals. The *chip-churr* appears to be a generalized preflight note signifying the readiness of the calling bird to take flight, for whatever reason. When a singing male was about to leave

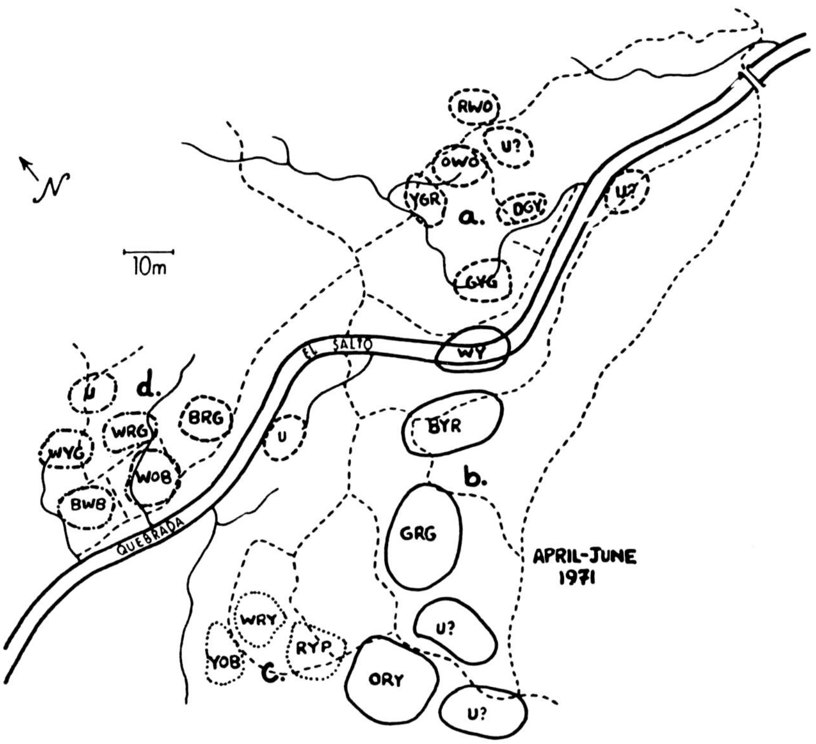
N
10m
RWO
U?
OWO
YGR
a.
DGY
U?
GYG
EL SALTO
WY
U
d.
BRG
WRG
WYG
WOB
BWB
U
BYR
b.
GRG
QUEBRADA
APRIL-JUNE
1971
WRY
RYP
U?
YOB
c.
ORY
U?

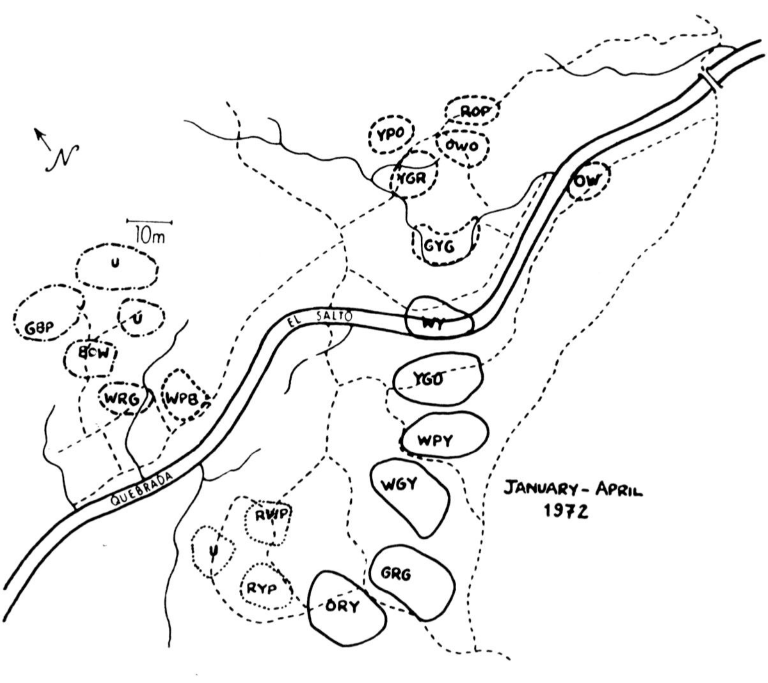
N
10m
ROP
YPO
OWO
YGR
OW
GYG
U
GBP
U
BCW
WRG
WPB
EL SALTO
WY
YGO
WPY
WGY
JANUARY-APRIL
1972
QUEBRADA
RWP
U
RYP
GRG
ORY

his territory in the absence of any intruder or challenger, he often mixed *chip-churrs* with his songs and frequently stretched his wings and/or tail and fluffed his feathers. The *chip-churr* appears to lack the aggressive character of the other preflight notes, the squeal and chatter.

Bill-pop.—A dry, snapping sound resembling a sharp, explosive kiss is called the bill-pop. It is produced simultaneously with a sudden snapping open of the bill (the gape display, described below). This sound was given by resident males in aggressive encounters at song perches, but was occasionally given by an intruding male as well. The social significance of this call is discussed more fully in connection with the gape display.

Visual Displays

In the absence of an intruder, a singing male sat upright on the perch with the wingtips held below the tail (Fig. 7A). Usually the head was moved through a small horizontal arc while the tail was wagged through a small vertical arc (less than 30°). Motion pictures indicate that neither head nor tail movements were timed to coincide with singing, and that one or both could be absent. The throat feathers were raised slightly giving the throat a ragged appearance and possibly accentuating the pattern of light and dark stripes.

When an intruder flew into or past a territory, the resident began singing louder and faster. Simultaneously the tail was spread with the two long central rectrices separated for most of their length, and was wagged vigorously through a vertical arc of up to 90°. At the apex of a wag, the tail formed nearly a right angle with the body. The throat feathers were very noticeably fluffed, and the crown feathers were often raised, accentuating the facial pattern (Fig. 7B).

The following displays (Fig. 7C–7I) occurred chiefly at the song perches of the resident males. Certain displays or components were given by lone birds, probably young males, at some distance from the lek.

In the float display, the displaying bird flies slowly back and forth over a distance of 7–15 cm in front of a perched bird. The flying bird moves parallel to the perch at the same level as, and facing the sitting bird. As it traverses this path, the bird pivots its body through an angle of approximately 75° to either side of the vertical. The bill is held at an angle well above the horizontal and is usually wide open, presenting the orange mouth lining and lower mandible as well as the throat pattern to the perching bird. The latter usually presents its bright orange gape to the hovering bird. The perched bird "tracks" the movements of the hovering bird, always keeping its bill pointed at the latter. This display was given most often by the resident male, but also was given by the intruder, particularly when it was another resident of the same lek.

Gape and bill-pop (Fig. 7D).—This display was given most often by the territory resident but also by the intruder, especially if it belonged to the same lek.

←

Fig. 6. Location of four song types on lek II in 1971 (above) and 1972 (below). Each song type (a–d) is indicated by a different symbolic outline of the territory of each resident (line, heavy dash, etc.). Despite the turnover of some resident individuals between years, the song type locations remained constant.

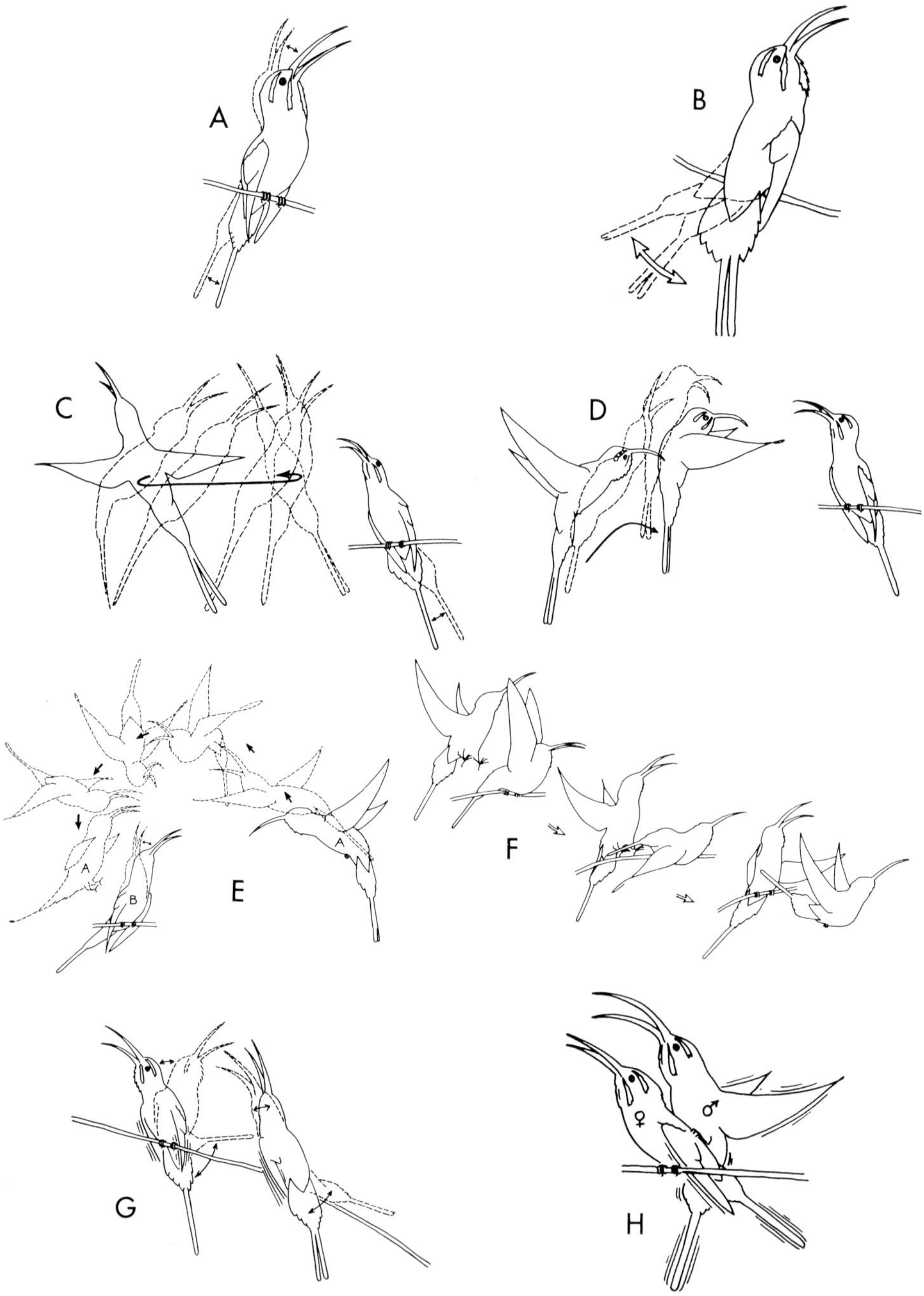

Fig. 7. Visual displays of *P. superciliosus* given around song perches in lek territories. A, posture of resident male singing in the absence of an intruder—the normal posture; B, resident male singing with vigorous tail wagging as an intruder approaches; C, float; D, gape and bill-pop; E, F, perch exchange sequence (F is a continuation of E as the flying bird settles toward the back of the perched bird); G, side-by-side; H, copulation. See text for descriptions of the displays.

The displaying bird hovers 15–20 cm from and facing the perched bird. The bird closes its bill and raises it to almost vertical as it starts a short, rapid dart in a shallow arc towards the perched bird. At the top of the arc the bird snaps its bill open and downward, flashing the bright orange gape and simultaneously producing the popping sound described earlier. The rami of the mandible appear to be spread slightly as the bill is opened, but much less conspicuously than in the corresponding *tock* display of *P. guy* (B. Snow 1974). As the bird reaches the closest point of the arc to the perched bird (ca. 10 cm away), the head is lowered and the bill closed. Several of these displays often were given in rapid succession, the displayer backing down and away to regain its original starting position. The gape and bill-pop was seen most often in aggressive encounters and frequently preceded or followed a perch exchange.

Perch exchange (Fig. 7E–F).—The perch exchange is usually initiated by the resident and is often given reciprocally by resident and intruder. After starting in a manner similar to a gape and bill-pop, the hovering bird circles over the sitting bird to approach it from above and behind. The circling bird has its bill open and pointed toward the perched bird; the throat feathers usually are fluffed. The circling bird settles down as though to land on the back of the perched bird, which flies from the perch, and the circling bird lands in its place. The formerly perched bird is now hovering in front of the perch, the positions of the two birds having been reversed. This perch exchange sequence was often repeated as many as 10 times in rapid succession; between perch exchanges the hovering bird might gape and bill-pop.

Side-by-side (Fig. 7G).—The hovering bird alights 12–15 cm from the perching bird, usually facing the opposite direction, or both birds land together following a hover-up. On the perch, each bird moves its head through a horizontal arc of approximately 60° toward and away from the other. The bills are wide open and held at an angle of 30° or so above the horizontal. The facial feathers are fluffed accentuating the pattern of dark and light stripes. The tails are usually fully spread and wagged vigorously through a vertical arc of 60–90°, and often from side to side as well. The wings are vibrated or fluttered below the body, and the back feathers may be raised. The high intensity side-by-side display appears in both aggressive and sexual contents and is seen less often than the preceding displays.

Hover-up.—Resident and intruder hover 15–25 cm apart, facing each other, as they rise together for a meter or more (depending partly on the configuration of the vegetation). The birds hold their bodies vertically; their bills are horizontal and usually closed. The posture resembles that of the bird that has just bill-popped in Figure 7D; in fact, one bird, usually (always?) the resident, may gape and bill-pop during the hover-up. Immediately following this display the resident usually darts slightly downward toward the intruder, chasing it away and pursuing it. Occasionally the birds hover up and down several times, sometimes finally perching in a side-by-side. The hover-up appears to be a high intensity aggressive interaction without sexual significance.

Chase.—Although not a display, chases are considered here because they are by far the commonest conclusion to any behavioral interaction between resident and intruder. The intruder flies rapidly from the territory, pursued by the resident for a few meters or, more often, for some distance. The chasing birds often flew back and forth around the lek, traversing the territories of other residents who sometimes joined in. Three- and four-bird chases usually originated as two-bird encounters. Very rarely did more than one intruder appear simultaneously in the territory of a resident.

It was often difficult to determine how a chase was initiated. When the chase started from a hover-up or when the intruder hovered while the resident perched, the resident's attack definitely started the chase. In many cases the two birds appeared to start off simultaneously, but perhaps there were subtle behavioral signals that we were unable to identify.

In addition to chase notes, we often heard a few song notes during a chase, particularly when both participants were lek residents. These songs seemed to occur most frequently when the chase passed through the territory of one resident or the other. Probably, a resident sang as the chase passed through his territory, but we were able to verify this only in a few cases where the two residents sang different dialects. It was usually impossible to identify the singer, or even to ascertain whether the same bird was consistently the chaser.

Phaethornis superciliosus appears to exhibit no clearly defined chain or sequence of displays although certain ones are sequentially associated (Table 3). For instance, a perch exchange tends to be followed by another perch exchange or by gape and bill-pop by the resident, depending upon which bird initiated the perch exchange. The course of an interaction often was determined largely by whether the intruder hovered before the resident or perched upon entering the territory. This in turn may have depended on the former's sex, age, and territorial status. Intruders that were residents of the same lek seemed much more likely to perform displays such as the float or gape and bill-pop upon entering the territory. We did not definitely observe females performing these displays.

The difference in position of certain displays in a sequence, depending on whether they were given by the intruder or the resident, may reflect changes in relative aggressiveness of the two birds. The aggressiveness of an intruding male who was challenging or testing the resident was highest at the start. During the succeeding interaction the aggressiveness of the intruder declined, culminating in a chase. The intruder should be least subordinate to the resident during the initial challenge, and most subordinate when he is chased. In a few cases the intruder, at the initial challenge, dominated the resident, who eventually lost control of part or all of his territory. This occurred most often when the resident was a young male that had just gained a territory or was still not in full control of one. Usually the dominant intruders were neighboring males (either adults or older, established young males), who then added the song perch(es) to their own territories.

Matings, True and False

In copulation one *P. superciliosus* mounts the other by alighting on its back following a gape and bill-pop, a float, or side-by-side (Fig. 7H). The mounted bird leans forward, its head held low and its back hunched, often fluttering its wings below the body. The mounting bird grasps the back feathers of the other

TABLE 3

OBSERVED TEMPORAL RELATIONSHIPS OF DISPLAYS IN 85 BEHAVIORAL INTERACTIONS IN *PHAETHORNIS SUPERCILIOSUS* LEK TERRITORIES[1]

Preceding event or display	Following event or display										
	Intruder gapes and bill-pops	Intruder floats	Perch exchange	Side-by-side	Hover-up	Resident gapes and bill-pops	Resident floats	Resident mounts	Chase	Copulation	Intruder leaves w/o being chased
Initiation of sequence											
Intruder enters and perches	–	–	3	7	1	4	13	–	2	–	–
Intruder enters and hovers	14	5	16	3	9	–	–	–	6	–	2
Intruder gapes and bill-pops	–	–	14	–	3	–	–	–	2	–	–
Intruder floats	–	–	5	1	–	–	–	–	2	–	–
Perch exchange	3	2	37	5	5	30	6	–	13	–	1
Side-by-side	3	–	5	1	3	–	3	2	4	–	2
Hover-up	–	–	1	–	–	2	–	–	23	–	–
Resident gapes and bill-pops	–	–	14	5	4	–	2	1	13	–	–
Resident floats	–	–	8	1	–	4	–	2	9	–	–
Resident mounts	–	–	1	–	–	–	–	–	1	3	–
Copulation	–	–	–	–	–	–	–	–	1	–	–

[1] Only those interactions that start with resident perched in his territory are included.

in its feet and vibrates its wings, presumably for balance. Both birds twist their tails to facilitate cloacal contact, and both vibrate their tails. On one occasion when the mounted bird was a female apparent cloacal contact lasted 3–5 sec; 2 other copulations between a male and an unsexed bird were of similar duration. In two homosexual mountings cloacal contact was much briefer if it occurred, and only the superior bird twisted its tail. The mounting and copulation positions in *P. superciliosus* are similar to those in other hummingbird species (B. Snow 1974; Wolf 1975a; Stiles and Ortiz-Crespo MS).

We rarely saw *P. superciliosus* mate. We witnessed only six probable copulations and two known homosexual mountings in several hundred hours of watching at lek territories. The copulations were immediately preceded by a float (twice), a perch exchange with the resident gaping and bill-popping (twice), and a side-by-side (twice). With one exception we have not seen a female perform any display to signal her sex to the resident. The exception occurred during a side-by-side when the intruder shuffled along the perch towards the resident male. The latter then performed a float as the former continued to fluff its feathers and wag its tail as in the side-by-side; the male mounted this bird and copulation followed.

Similarly, we saw no special male precopulatory display distinct from the aggressive displays occurring in male-male encounters. At least some of these displays, e.g., the perch exchange, are performed by the female as well. We never saw a female give the float or gape and bill-pop. The general copulation sequence thus differed rather little from that of many male-male aggressive encounters until the actual mounting. We unfortunately were unable to film a copulation sequence, and thus may have missed subtle behavioral differences from a male-male encounter. However, the mountings following perch exchanges seemed to occur simply because the sitting bird remained on the perch and allowed the circling bird to land on its back; the latter then immediately attempted copulation. A female's major means of signalling her sex and receptivity may be just to stay perched during such a sequence; were she to fly, she probably would be chased. Conversely, a male *P. superciliosus* may treat as a female, and attempt to copulate with, any bird that does not fly as he circles behind it.

The two homosexual mountings support this interpretation. As we marked our first *P. superciliosus*, we inadvertently pulled out most of its tail. When released, this bird (YY) flew off approximately 6 m and preened and fluffed its plumage. Almost immediately another *P. superciliosus* floated in front of YY, who seemed to ignore it. The second bird circled over and behind YY and landed on its back in what appeared to be an attempted copulation, at which point YY flew off. YY turned out to be a resident male on the lek (Ia) with a territory only 5 m from the site of the attempted copulation. However, because YY did not fly during what appeared to be a perch exchange, he was treated as a female by the other bird.

The second homosexual mounting involved a marked resident of lek Ic (ROY) and a marked bird that turned out to be a nonresident male (BOP). BOP failed to fly during a perch exchange, and ROY alighted on its back and attempted copulation. After a second or two, BOP shook loose and flew off alone.

Males frequently attempted copulations with dead leaves about hermit size and suspended by a spider web. This behavior was most often shown by adult males on their territories, but we occasionally saw intruders calling from a territory in the resident's absence engage in leaf copulations. We also saw this behavior by several individuals well away from any lek, once by a young, nonresident male. In a typical leaf copulation the male flies slowly toward the leaf, which is usually 2–3 m from the perch and somewhat above it. He then gapes and bill-pops several times at the leaf, and may float before hovering in a vertical position with his feet grasping the leaf and his abdomen in contact with it. At contact, the abdomen and tail are vibrated in a manner similar to true copulation. Following the leaf copulation, the male usually flies backward, gapes and bill-pops again, and flies off; occasionally he repeats the performance up to five times before leaving. Checks of several leaves revealed no trace of fluid. Similar behavior, called "leaf mating" or "false mating," has been reported in several other species of hermit hummingbirds (Arp 1957; B. Snow 1973a, b, 1974; pers. obs.).

All true matings and homosexual mountings that we saw occurred on leks, usually within the territory of the male. Four copulation sequences and one homosexual mounting occurred at song perches of the mounting male. In one instance, the presumed female flew into the territory and perched about 3 m from the song perch of the male, who approached and copulated with her there. In the one copulation that occurred outside of a male's territory, the initial part of the sequence occurred within the territory; it was followed by a chase of approximately 20 m, at which point the female perched, and the final displays and mating took place. The one homosexual mounting occurring outside a territory, that involving YY (see above), may have involved a neighboring male that left his territory on seeing YY perched. These observations tend to confirm the importance of lek territories as mating stations. All matings occurred between 0730 and 0930 although our observations on lek territories covered the entire day. The small number of matings observed could result from three causes: (1) the number of matings is, in fact, low; (2) many matings occur away from the lek; and (3) the dense vegetation blocked our view of some matings. Probably all three factors contribute. We present evidence later that during the main lekking season only about one female visits a 15-male lek each day to mate. Thus, an observer at a randomly chosen territory would have, at best, a 6.7% chance of seeing mating during a day's observations (assuming he could see all parts of the territory, which is not always the case). One mating away from a male's territory began as a chase from within the territory; this may be advantageous to the male by reducing interference from other lek residents during mating. Given the rarity with which females appear on the lek, a male may stand a better chance of mating if he chases a potential female than if he lets her go and waits for another to appear. Similarly, nonmating chases should have little effect on male presence when the next receptive female arrives, a very rare event.

FORAGING PATTERNS

Food Plants

We have records of *P. superciliosus* foraging at the flowers of 21 species of plants (Table 4). Doubtless other species are visited occasionally, but these data should give a fairly complete picture of flower use by the La Selva population.

TABLE 4

MORPHOLOGY, PERIOD OF GOOD BLOOM, AND HUMMINGBIRD VISITATION OF FOOD PLANTS OF *PHAETHORNIS SUPERCILIOSUS*

Food plant	Mean effective corolla length (mm)[1]	Corolla shape[2]	Months in good bloom[3]	Proportion of total observed visits by: Hermits	Non-territorial nonhermits	Territorial nonhermits	Daily nectar production (μl/flower)
Heliconia pogonantha (Musaceae)	33	C	Dec–May*	0.69	0.10	0.21	115
H. wagneriana (Musaceae)	48	C	Feb–Apr	0.98	0.02	–	90
H. mariae (Musaceae)	22	SC	Feb–Jun*	0.30	0.51	0.19	45
H. imbricata (Musaceae)	21	SC	Jun–Sep	0.15	0.26	0.59	85
H. latispatha (Musaceae)	32	S	Jun–Sep	0.09	0.32	0.59	75
H-3[4] (Musaceae)	41	C	Aug–Oct*	0.94	0.06	–	70
H-18[4] (Musaceae)	44	C	Jul–Sep	0.99	0.01	–	65
H-16[4] (Musaceae)	40	SC	Jul–Aug	1.00	–	–	70
H-17[4] (Musaceae)	28	S	Aug–Oct	0.26	0.29	0.45	110
Calathea lutea (Marantaceae)	33	C	Feb–May	1.00	–	–	20?
Costus ruber (Zingiberaceae)	35	C	May–Jul	0.97	0.03	–	85
C. malortieana (Zingiberaceae)	40	C	Aug–Nov	0.96	0.04	–	75
Aechmea magdalenae (Bromeliaceae)	26	SC	Jul–Aug	0.89	0.11	–	—
Passiflora vitifolia (Passifloraceae)	30	(S)	Feb–Mar	1.00	–	–	85+
Gurania levyana (Curcurbitaceae)	17	S	Feb–Mar	0.11	0.69	0.20	—
Malvaviscus arborea (Malvaceae)	38	S	Feb–Apr, Oct*	0.87	0.13	–	80
Pentagonia donnellsmithii (Rubiaceae)	34	S	Mar–Jun	0.92	0.08	–	30
Jacobinia umbrosa (Acanthaceae)	28	C	Feb–Mar	0.57	0.40	0.03	70
Aphelandra storkii (Acanthaceae)	36	C	Aug–Oct	0.91	0.09	–	55
Columnea nicaraguensis (*Gesneriaceae*)	25	SC	Apr–Jun, Oct–Nov	0.08	0.49	0.43	—
Alloplectus coriaceus (Gesneriaceae)	25	SC	Sep–Nov	0.24	0.57	0.19	—

[1] Minimum distance from the mouth of the corolla tube to the nectar chamber.

[2] C = curved, SC = slightly curved, S = straight. For *Passiflora* the tube of the corolla is broad and essentially straight, but arrangement of the stamens effectively prevents any but a long-billed hummer from entering the flower.

[3] * = flower that blooms year round.

[4] Taxon numbers of presently undescribed *Heliconia* species, as used in a revision of the genus (Daniels and Stiles 1979; see also Stiles 1975). [See note p. 78.—Ed.]

TABLE 5

FEEDING RECORDS OF *PHAETHORNIS SUPERCILIOSUS* AT LA SELVA[1]

Food plant	J	F	M	A	M	J	J	A	S	O	N	D	Total	Proportion
Heliconia pogonantha	25	40	39	24	16	16	35	12	7	6	17	18	265	0.27
H. wagneriana	2	16	28	20	2	–	–	–	–	–	–	–	68	0.07
H. mariae	1	–	1	4	4	2	1	–	–	–	–	1	13	0.01
H. imbricata	–	–	–	–	2	21	31	21	9	2	–	–	86	0.09
H. latispatha	–	–	–	1	–	5	4	2	–	–	–	–	12	0.01
H-3	6	7	5	3	4	5	16	13	12	7	3	4	85	0.09
H-18	–	–	–	–	–	1	29	24	13	7	–	–	74	0.08
H-16	–	–	–	–	–	1	4	11	2	–	–	–	18	0.02
H-17	–	–	–	–	–	–	–	4	7	6	–	–	17	0.02
Calathea lutea	–	1	1	2	–	–	–	–	–	–	–	–	4	0.004
Costus ruber	–	–	3	12	14	15	7	1	2	–	–	–	54	0.06
C. malortieana	6	1	2	2	–	4	12	8	10	13	11	4	74	0.08
Aechmea magdalenae	–	–	–	–	–	2	5	2	5	–	–	–	14	0.01
Passiflora vitifolia	1	9	8	4	–	–	–	–	1	3	–	1	27	0.04
Gurania levyana	–	2	4	2	1	–	–	–	–	–	–	–	9	0.01
Pentagonia donnell-smithii	2	2	4	6	4	3	1	–	2	1	2	1	28	0.03
Malvaviscus arborea	4	5	7	6	6	1	2	1	3	2	2	3	42	0.04
Jacobinia umbrosa	5	11	13	6	–	–	2	–	–	–	–	–	39	0.04
Aphelandra storkii	–	–	–	–	–	–	2	10	12	4	1	–	29	0.03
Columnea nicaraguensis	–	–	1	1	–	–	–	–	2	3	1	–	8	0.01
Alloplectus coriaceus	1	1	–	–	–	–	–	1	2	–	1	–	5	0.005
No. feeding observations	52	95	116	93	53	76	151	111	89	53	38	44	971	
No. food plants visited	10	11	13	14	9	12	14	13	15	11	8	7		

[1] Obtained during census walks and general observations, 1969–1973; not including observations at specific flowers.

Of the 21 species, 14 are visited primarily by hermit hummingbirds, which in most cases comprise more than 90% of the visitors (Table 4).

Feeding records of *P. superciliosus* at these 21 plant species are given in Table 5. These data can be used to divide the plant species into four use categories, as follows: a species that contributes (1) less than 10% of foraging records in all months, and less than 2.5% overall; (2) at least 10% of foraging records in 1 or more months, and/or 2.5% overall; (3) at least 20% of foraging records in 1 or more months and/or 5% overall; (4) at least 40% of foraging records in 1 or more months and/or 10% overall.

Heliconia pogonantha is the only species in category 4 and was by far the most important food plant of *P. superciliosus* at La Selva. It had the highest nectar production of any heavily used flower (Table 4) and was the most frequently visited for 8 months. It also had the longest period of good bloom of any major food plant (December to May or June in most years), due largely to its production of large, long-lived inflorescences (Stiles 1975).

Use category 3 includes four other species of *Heliconia* and two of *Costus*. *Heliconia* H-3 was of secondary importance in any one month, but bloomed and was visited year round. The other species had short, well-defined periods of good

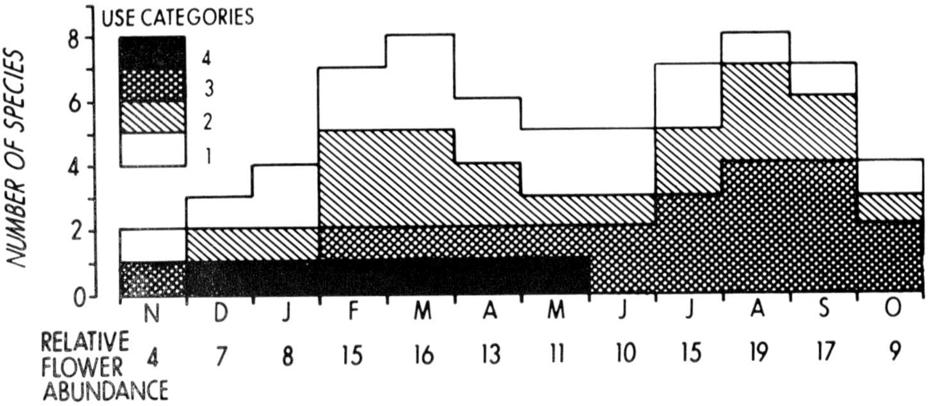

Fig. 8. Number of species of food plants of *P. superciliosus* in good bloom in different months, according to use categories (see text). The relative flower abundance is obtained by multiplying the number of each use category by the number of plant species in good bloom in that category (e.g., for February: $4 \times 1 + 3 \times 1 + 2 \times 3 + 1 \times 2 = 15$).

bloom during which they were visited frequently, and long periods of little or no blooming or visitation (Tables 4, 5). *H. wagneriana* was important during the dry season, *H. imbricata*, H-18, and *Costus ruber* during the early wet season, and *C. malortieana* during the mid- to late wet season. Thus, all of the major food plants of *P. superciliosus* at La Selva were large herbaceous monocots in only two genera.

Category 2 includes species that are visited regularly but not frequently. These plants generally are uncommon or very locally distributed and tend to have relatively few flowers per plant and/or little nectar per flower. H-17 is a nectar-rich species often visited and defended by nonhermits, which reduced hermit usage. The flowers in category 1 were visited only occasionally by most *P. superciliosus*, although certain individuals visited these species regularly. In contrast to categories 3 and 4, categories 1 and 2 contain a considerable taxonomic diversity with representatives of 9 families, including 3 of monocots and 6 of dicots. *Malvaviscus* was the only common species in category 1 or 2 with a long blooming season. Some individuals could be found in bloom every month, but the population rarely flowered synchronously. The other minor food plants had relatively short, well-defined flowering seasons, either in the dry season (6 species) or the early to mid-wet season (6 species) or both (*Columnea nicaraguensis*). *Heliconia mariae* is known from only two clumps at La Selva; it bloomed year round, but with a strong peak in the late dry and early wet seasons (Stiles 1975, 1978a).

The use categories can be used to construct a semiquantitative weighting system that allows us to estimate the overall flower abundance and how this changes seasonally (Fig. 8). There were definite peaks of flower availability in the mid-dry season (February through March or April) and the early wet season (July through September). A short but very pronounced lean season occurred between late October and early December in most years. The only food plant of *P. superciliosus* in category 3 or 4 in good bloom during this period was *Costus malortieana*, which was locally common in second-growth and overgrown cacao in the northern quarter of the property but very rare elsewhere. Thus, the *P. superciliosus* populations of the southern part of the property either had to move into the northern part or subsist mostly upon insects. We have several records of changes of feeding areas by individuals at this season. For example, BWY (a female by measurements) was consistently recaptured or observed during the months December–August of 1971 and 1972 only along the east boundary where *Heliconia pogonantha* was abundant; but during September–November, this bird was caught or observed only in large *Costus* patches north and west of lek I. Similarly GBP, a resident male on lek II, was recaptured in the *Costus* patches in November 1972; other feeding records for this bird are for the east boundary. The period of reduced nectar availability ended in early to mid-December when *Heliconia pogonantha* attained good bloom. However, the lengths of the lean season and the blooming seasons of most food plants can vary by a month or more in different years (Stiles 1978a).

Data on the daily pattern of nectar production are available for all the *Heliconia* species (Fig. 9A, Stiles 1975) and several other food plants (Fig. 9B). Nectar production was highest in the early morning and declined to low values by early afternoon. Nectar harvested by hermits in the early morning was replaced, but by midday nectar was a nonrenewable resource, at least until the next day. Nectar harvested in the afternoon consisted mainly of that not taken during the morning. Flowers of all these species lasted only one day (Stiles 1975, pers. obs.) necessitating some day-to-day shifts in foraging. The foraging patterns (and probably other activities as well) of *P. superciliosus* must reflect hourly, daily, and seasonal variations in nectar availability.

The weighting system based on use categories can also be used to evaluate certain characteristics of the food plants (Fig. 10). Apparently *P. superciliosus* visits mainly plants whose flowers have long, curved corollas and that are relatively little used by territorial nonhermits. B. Snow and D. Snow (1972) and Stiles (1975) showed that the hummingbirds in two lowland tropical areas visited flowers falling into two groups based on corolla morphology. Species with long, curved corollas were visited mainly by hermits (genera *Eutoxeres*, *Phaethornis*, *Threnetes*, *Glaucis*) that have curved bills 30 mm or more in length. Most nonhermit hummingbirds have straight bills 25 mm or less in length, and tend to visit flowers with short, straight corollas. Among the major food plants of *P. superciliosus*, only *Heliconia imbricata* was visited primarily by nonhermits. Nonhermits were the major visitors to several food plants in categories 1 and 2.

Interspecies Relations

An important behavioral difference between nonhermits and hermits is that the former (mostly males) frequently hold flowered-centered territories while the latter do so very rarely and inconsistently (B. Snow and D. Snow 1972; Linhart 1973; Stiles 1975). The territories of nonhermits at flowers are primarily for feeding, although they may serve secondarily as mating sites (pers. obs.; see also Pitelka 1942). The critical resource being defended is the flowers, and they are defended against all other hummingbirds, regardless of species or sex. At La Selva, the nonhermits that held territories at species of plants used by *P. superciliosus*

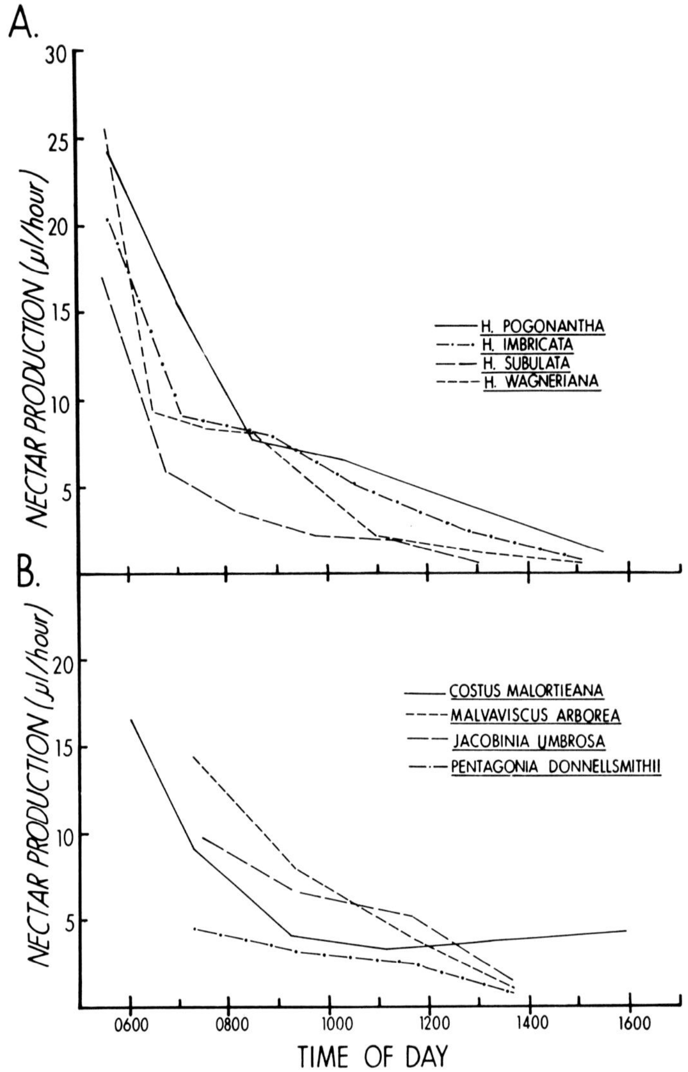

Fig. 9. Daily nectar production of eight plant species frequently visited by *P. superciliosus*. A, four species in the genus *Heliconia*; B, four non-*Heliconia* species. [See note p. 78.—Ed.]

included males of *Chalybura urochrysia* (7 g), *Thalurania furcata* (4.5 g), occasionally *Florisuga mellivora* (6.5 g), and *Amazilia amabilis* (4.2 g), and possibly both sexes of *Amazilia tzacatl* (5 g). Females of these bird species usually were nonterritorial and often visited different flowers from males, or poached nectar from defended flowers. These females, plus individuals of several smaller species of hummingbird that occasionally visited the food plants of *P. superciliosus* (e.g., *Klais guimeti*, 3 g) are grouped as "nonterritorial nonhermits" in the following discussion.

When two hummingbirds meet at a clump of flowers, they almost always interact aggressively; normally one bird forces the other to move away or leave the flowers

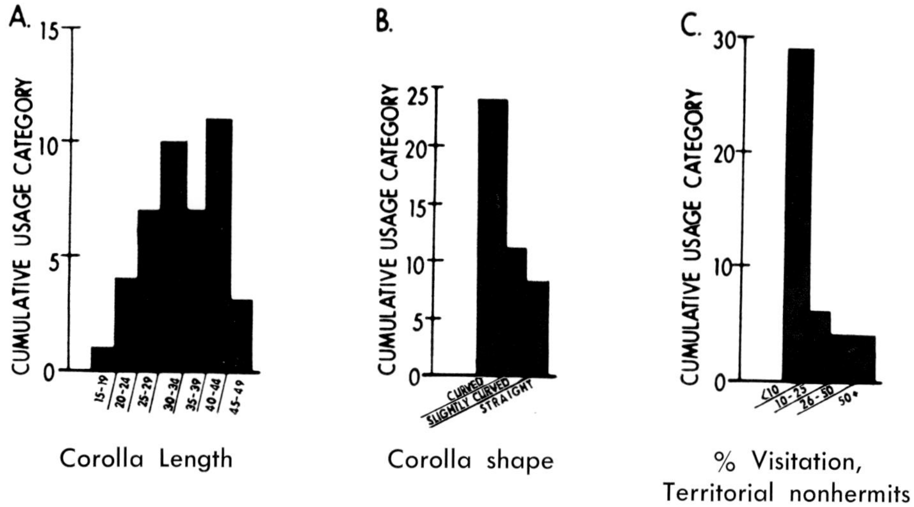

Fig. 10. Cumulative use categories of food plants in relation to: A. corolla length, B. corolla shape, and C. percent of visits to a food plant by territorial nonhermits. Note changes in the ordinal scale.

entirely. Tables 6 and 7 present the outcomes of such aggressive interactions involving *P. superciliosus* at different flowers. *P. superciliosus* almost invariably lost in any such aggressive interaction if the other bird was a nonhermit, regardless of its species, sex, or weight or the food plant at which the interaction occurred (Table 6). Territorial nonhermits always drove *P. superciliosus* away from the flowers; nonterritorial nonhermits almost always forced *P. superciliosus* to give way, but rarely actually chased them.

Most interactions with territorial nonhermits occurred at *Heliconia pogonantha* and *H. imbricata*. The latter is adapted for pollination by nonhermits (Stiles 1975), and large clumps regularly were defended by male *Chalybura*, *Thalurania*, and *A. tzacatl*. *P. superciliosus* tended to visit mostly small, undefended clumps of *H. imbricata*, or tried to feed at the edges of larger, defended clumps. The flowering phenology and corolla morphology of *H. pogonantha*, in contrast, are specialized for hermit pollination. However, the largest of the nonhermits, *Chalybura*, has a bill just long enough (22–25 cm) to negotiate the corolla of *H. pogonantha* although it could not forage very efficiently at this flower (pers. obs.). Where several large clumps of *H. pogonantha* grew close together, male *Chalybura* frequently were territorial. This occurred mainly in large areas of second growth and largely resulted from human disturbance. In undisturbed habitat, as along forest streams, most clumps of *H. pogonantha* were too small to be defended by *Chalybura*, and far fewer interactions occurred (Stiles 1975). If the nonhermit was at all consistent in its territorial defense the amount of nectar a hermit could obtain at the defended clump might be too low to justify the time and energy expenditures to visit it. The result would be to effectively remove large clumps from the floral food resources available to *P. superciliosus*.

By contrast, dominance interactions appeared to play at most a minor role in

TABLE 6

DOMINANCE RELATIONS OF *PHAETHORNIS SUPERCILIOSUS* AT SEVERAL FOOD PLANTS

Food plant	Hrs. of observation	Encounters with other hermits		Encounters with nonterritorial nonhermits		Encounters with territorial nonhermits	
		Won[1]	Lost	Won	Lost	Won	Lost
Heliconia pogonantha	68	7	10	1	7	0	46
H. wagneriana	26	3	10	0	0	0	0
H. latispatha	29	0	0	0	1	0	7
H. imbricata	49	1	1	0	8	0	59
H. mariae	8	1	0	0	1	0	3
H-3	14	4	1	0	0	0	0
H-18	27	4	0	0	1	0	0
H-16	10	1	0	0	0	0	0
H-17	23	0	0	0	3	0	12
Costus ruber	7	1	0	0	0	0	0
C. malortieana	19	5	4	1	6	0	0
Malvaviscus arborea	5	0	0	0	1	0	0
Jacobinia umbrosa	10	3	1	0	2	0	1
Totals		30	27	2	30	0	128

[1] Won indicates that *P. superciliosus* was dominant in the encounter; lost means that *P. superciliosus* was subordinate.

determining feeding patterns at flowers visited mostly or exclusively by hermits. Interactions between hermits were much less frequent than those involving nonhermits, in part because hermits were much less aggressive even when two birds arrived simultaneously at a clump. Weight played a role in these interactions as *P. superciliosus* was dominated by *Eutoxeres* (10–12 g) but dominated *P. longuemareus* (2.5 g). No consistent pattern of dominance relations was evident between *P. superciliosus*, *Threnetes ruckeri* (5.6 g), and *Glaucis aenea* (5.3 g) (Table 7). The former tended to be dominant at *H. pogonantha* and H-18, and subordinate to *Glaucis* at *H. wagneriana*. This was perhaps due to habitat because *Glaucis* often congregated and nested in the open second growth areas where *H. wagneriana* grew. This species and its close relative in Trinidad, *G. hirsuta*, defend the nesting area extremely vigorously, often attacking other hummingbirds at some distance from the nest itself (Skutch 1964a; B. Snow 1973a; Stiles, pers. obs.).

STRATEGIES OF FLOWER EXPLOITATION

Depending on the spatial distribution of flowers, several alternative strategies of nectar harvesting are available to hummingbirds (Janzen 1971; Hainsworth and Wolf 1972a; Linhart 1973; Stiles 1975). Territoriality requires a rich, localized nectar supply so that the costs of defense are outweighed by the advantages of exclusive access (Brown 1964; Wolf 1969; Gill and Wolf 1975; Wolf et al. 1975). Commuting occurs when a bird defends a nonfood-centered territory at one site and travels regularly to another site to feed (Stiles 1973). There may be no single feeding area but rather a series of small, scattered feeding sites that are arranged along a more or less regular route or "trapline." These strategies are not mutually exclusive but represent points along a continuum from clumped to dispersed nectar sources, from entirely within to entirely outside the territory.

TABLE 7

DOMINANCE RELATIONS OF *PHAETHORNIS SUPERCILIOSUS* AND OTHER HERMITS AT DIFFERENT FOOD PLANTS

Food plant	Encounters with *Eutoxeres aquila*		Encounters with *Threnetes ruckeri*		Encounters with *Glaucis hirsuta*		Encounters with *Phaethornis longuemareus*	
	Won[1]	Lost	Won	Lost	Won	Lost	Won	Lost
Heliconia pogonantha	0	7	2	1	5	2	–	–
H. wagneriana	–	–	1	0	2	10	–	–
H-3	–	–	–	–	1	0	3	1?
H-18	–	–	1	0	3	0	–	–
Costus malortieana	–	–	3	3	0	1	2	0
Jacobinia umbrosa	–	–	–	–	0	1	3	0
Others	–	–	0	0	1	1	2	0

[1] As in footnote 1, Table 6.

All available evidence indicates that *P. superciliosus* employs only commuting and regular foraging routes in flower visitation. We have observed consistent territorial defense only at the lek (males) or at the nest (females), never in connection with flowers. Territorial nonhermits exclude *P. superciliosus* from most clumps of flowers large enough to be economically defensible. Moreover, flowers specialized for hermit pollination rarely provide enough nectar at a single time and place to make territoriality feasible (Stiles 1975). Thus, some movement between nectar sources is obligatory. With small and scattered nectar sources, foraging is most efficient if flower clumps are visited in a regular order so that reward levels are maximized and the time and energy expended in moving between clumps are minimized (Schoener 1971; Wolf et al. 1975). Most of our evidence for use of this foraging strategy by *P. superciliosus* is indirect. Due to the density of the vegetation and the distances involved, we have been unable to trace any foraging route in its entirety. However, we have observed a number of the essential features of such an exploitation system: (1) A given clump of flowers was visited by a given individual at regular intervals through the day, over a period of days. (2) The directions of arrival at and departure from a given clump remained fairly constant over many visits. (3) Any given clump was visited by one or a very few individuals, with rather little overlap between individuals in flowers used. If a hermit visited a clump sufficiently large that all its nectar could not be harvested in a single bout of continuous feeding, it often remained in the vicinity of the clump for as long as 15–30 min. During this time, it sometimes chattered at or even attacked other hermits attempting to feed at the clump. This was the closest *P. superciliosus* came to feeding territoriality, and it occurred only at clumps not visited or defended by nonhermits.

Certain food plants of *P. superciliosus* reached very high densities in second growth areas, and many lek males commuted to these areas to feed. The major feeding areas at La Selva included (see Fig. 1): the Point (chiefly *H. imbricata* and *H. wagneriana*), the East Boundary (chiefly *H. pogonantha*), and the old cacao and second growth between lek I and the field station (*H. imbricata*, *H. pogonantha*, and *Costus malortieana*). Once it arrived at a feeding area, *P. superciliosus* often visited one flower clump and then passed by several others before visiting a second. During regular censuses in the East Boundary area on several days in late 1971 and early 1972, we located four *H. pogonantha* clumps that were used consistently and probably in sequence by one *P. superciliosus* (ROW). Distances between clumps were approximately 24, 60, and 85 m, and intervening clumps were not visited. ROW probably was a female (by measurements), but similar data are available for a known lek male, GPB. This bird always emerged from the forest at the same point, visited one large clump (often remaining 5–10 min), flew directly to visit another clump approximately 80 m away, and then continued on beyond the census area. A mist net set across the path would consistently catch the bird. The route remained stable for over one year.

The spatial distribution of foraging flights of lek males paralleled that of floral food resources around the lek. The birds of lek I usually foraged within 250 m of the lek, mostly to the north (Fig. 11A). A few birds also commuted to feeding areas along the Quebrada Arbolera or up the Quebrada El Surá; at certain seasons a few individuals commuted over 1 km to the Point or the East Boundary. The East Boundary, though 0.5 km away, was the closest large feeding area to lek III (Fig. 11B).

Clumps of flowers (e.g., H-16, H-17, H-18) along a trail midway between leks I and II were observed intensively in August–September 1972 and July–August 1973 for visitation by marked birds (nearly all males on these leks were marked). Of the 48 observed foraging bouts 46 were by unmarked individuals, probably mostly females. Six of 7 *P. superciliosus* (all unmarked) mist-netted along this trail in 1972 were clearly females. During July 1973, the trail between lek I and the field station was used as a transect to determine the number of flower visits made by lek males at different distances from the lek. A majority of flower visits was by known lek I males at distances of up to 300 m from the lek; at distances greater than this, most visits were by unmarked birds and females (Table 8).

These observations suggest that resident male *P. superciliosus* fed reasonably near the leks, while females and perhaps nonlek males used areas some distance away. This assumes that the observed sex ratio of flower visitors in an area is representative of the sex ratio of the population present in that area. A bird will probably include in its regular foraging itinerary those flower clumps at which it harvests enough nectar to make revisitation profitable. These are the clumps that have not been visited recently by another hummingbird. By visiting a clump and removing its nectar, a *P. superciliosus* makes visitation unprofitable for a considerable time thereafter, which should make it unlikely that another bird will visit the clump regularly. In this way such an exploitation system can produce exclusive or preponderant use of at least small clumps by the same individual. Given the concentration of males about leks, females must forage mainly at some distance from leks to encounter sufficient unvisited clumps to make an energetic profit. This may explain why most *P. superciliosus* nests we found were well away from leks (Fig. 1).

Nectar vs. Insects

The diet of *P. superciliosus* includes small arthropods as well as nectar, but the relative proportions of each are difficult to determine due to the great differences

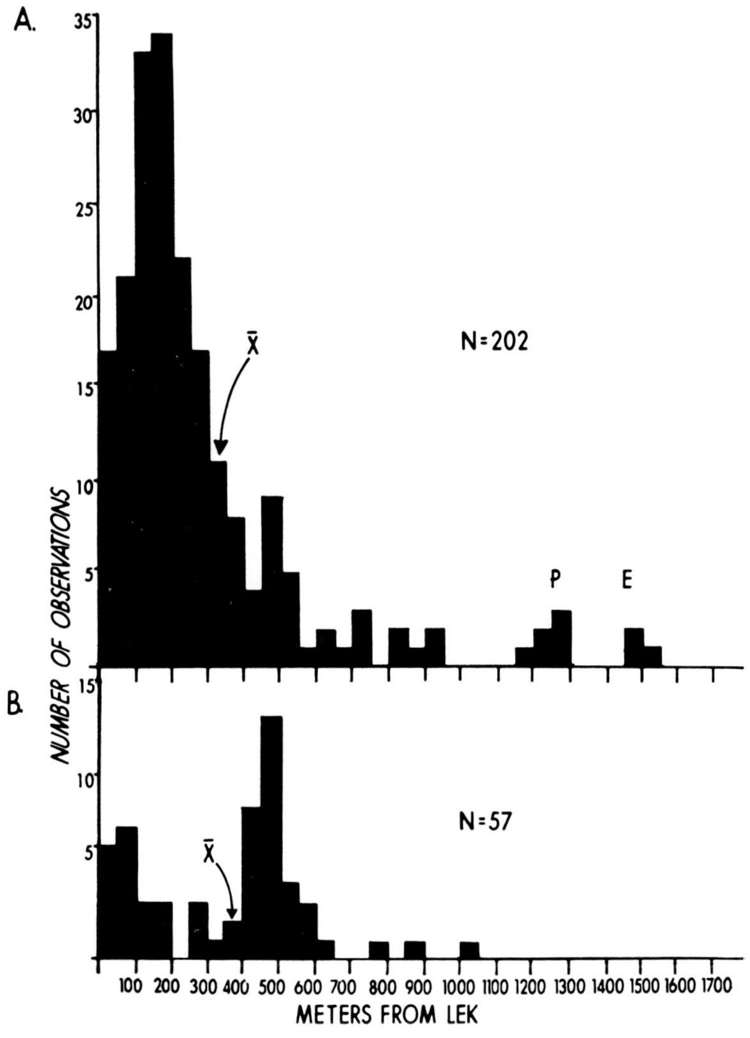

Fig. 11. Records of distances at which individual male *P. superciliosus* foraged from their territories. A. lek I; B. lek III. We excluded records from long-term observations in major feeding areas. P = records from the Point feeding area; E = records from the east boundary.

in foraging techniques required. Examinations of stomach contents are not helpful because of the extremely rapid postmortem digestion of nectar. The arthropod food of *P. superciliosus* consists almost entirely of spiders gleaned from foliage, dead twigs, or webs in the mid- to upper understory of forest, or the canopy of young second growth (Stiles and Hespenheide MS). *P. superciliosus* foliage-glean usually well away from lek territories or flowers, often along streams or at breaks in the forest canopy.

Compared with nectar, spiders should represent a food source the absolute availability of which changes relatively little through the day; indeed foliage-gleaning activity was much more evenly distributed over the day than was flower visitation (Table 9). An overall reduction in activity through the day is indicated

TABLE 8

OBSERVED FLOWER VISITS BY LEK MALES AND OTHER INDIVIDUALS ALONG A 450-M TRANSECT BETWEEN LEK I AND THE FIELD STATION

Distance from Lek I	Visits by Known lek ♂♂	♀♀ and unsexed birds marked at leks	Unmarked birds
0–150 m	18	7	2
150–300 m	9	1	6
300–450 m	5	4	11
450+ m	1	0	3
Totals	33	12	22

by the decline in total encounters, but the proportion of birds seen foliage-gleaning actually increased through the day.

TEMPORAL PATTERNING OF LEK ACTIVITY

BREEDING AND LEKKING SEASONS

Our limited sample of 15 nests of *P. superciliosus* from La Selva indicates that eggs were laid between January and August with peaks in January–March and June (Fig. 12C). At a nearby locality, females with enlarged gonads were collected between early December and September, but not in October or November (Fig. 12A). Adult males on the other hand had enlarged testes year round (Fig. 12B).

Consideration of the timing of molt in the population and the presence of bill corrugations on young birds (Ortiz-Crespo 1972) allows more detailed and quantitative evaluation of the nesting season. We have shown previously that individual *P. superciliosus* molt at almost exactly 12-month intervals, commencing their first molt about 12 months after hatching. For a bird in any stage of primary molt, we can estimate the date on which molt started; this should be close to the

TABLE 9

DAILY ACTIVITY PATTERNS OF *PHAETHORNIS SUPERCILIOSUS* IN DIFFERENT HABITATS

Habitat	Time of day	Total no. contacts[1]	Feeding at flowers	Foliage gleaning
Forest[2]				
	Early morning (0600–0900)	54	15	7
	Midday (1000–1300)	27	7	5
	Afternoon (1400–1700)	18	2	4
Second growth, with *Heliconia pogonantha*[3]				
	Early morning (0600–0900)	85	33	4
	Midday (1000–1300)	72	24	5
	Afternoon (1400–1700)	39	15	3

[1] Total no. contacts = total number of times a *Phaethornis* was seen or heard on census.
[2] Ca. 20 census hr for each time of day, over all times of year.
[3] Ca. 15 census hr for each time of day, all times of year.

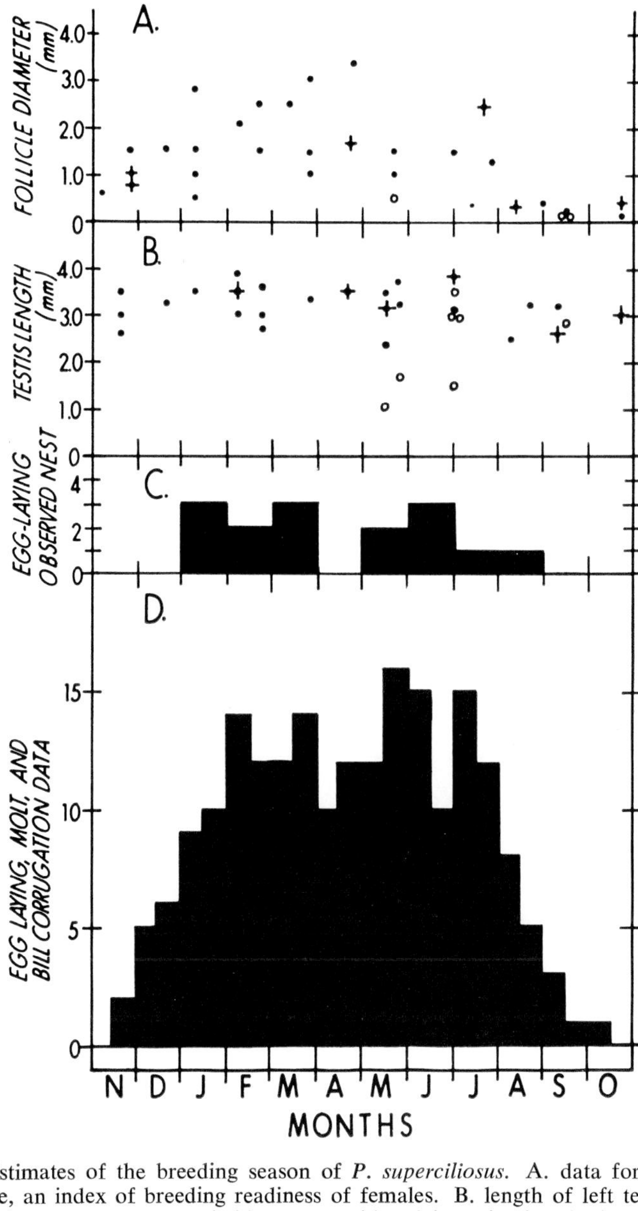

Fig. 12. Estimates of the breeding season of *P. superciliosus*. A. data for size of largest ovarian follicle, an index of breeding readiness of females. B. length of left testis; males with testes more than 2.5 mm long probably are capable of inseminating females. In A and B, closed circles are adults, open circles are juveniles, and crosses through a circle indicate a molting bird. C. estimated dates of egg-laying in nests observed at La Selva, assuming a 21-day incubation period. D. dates of egg-laying estimated from timing of molt of adults and the bill corrugation characteristics of immatures. See text for details of this method.

calendar date of hatching. We have also shown that bill corrugations can be used to age young birds up to about 6 months (Stiles and Wolf 1974). If we subtract three weeks for the incubation period of *P. superciliosus* (Skutch 1964a), we obtain the approximate date on which the egg from which the bird hatched was

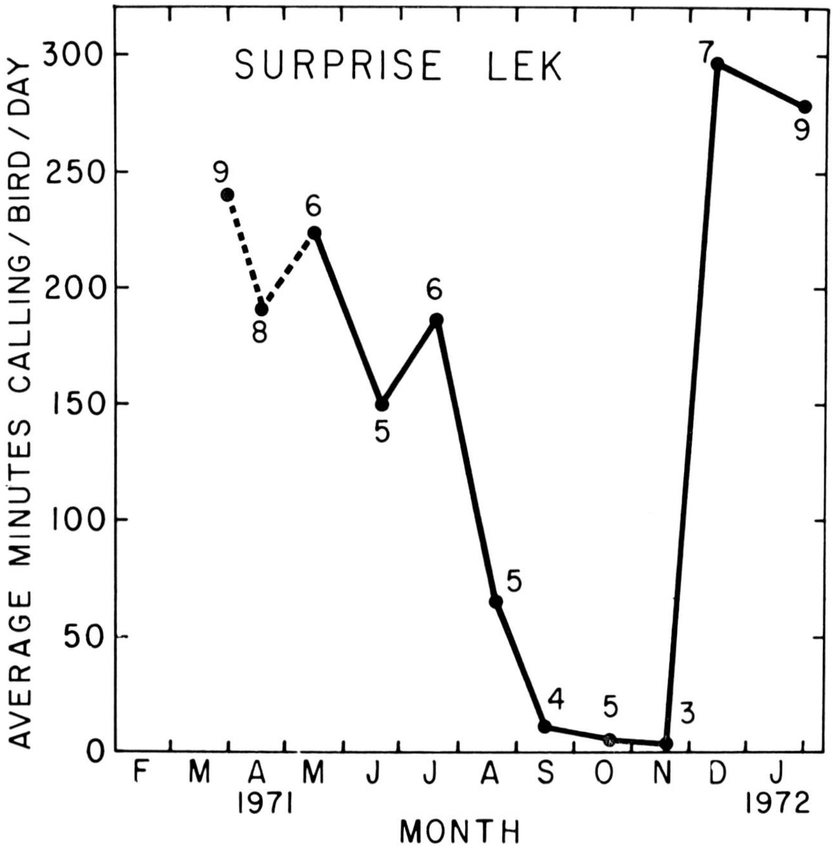

Fig. 13. Annual variation in calling activity among resident males on lek Ic (Surprise lek) during one nearly complete annual cycle of lek occupancy. Numbers beside each point indicate number of males for which data were recorded. The dotted line indicates that the records for April are for only one half-day; the others are for full days.

laid. As this method estimates the dates of egg-laying in successful nests only, any marked seasonal variations in nest success will bias this estimate of the nesting season.

We have estimated the nesting season from data on molt timing and/or bill corrugations for 208 individuals. It agrees fairly well with estimates from nest records and gonad data (Fig. 12D). Apparently nesting occurred in every month of the year but was very rare in September, October, and November. The main nesting season was January–July, with a sustained high level of nesting from February onward. This represents the average nesting season over several years, but it may vary considerably from year to year as birds of different ages contribute to the estimate.

The best available measure of seasonal variation in lek activity is given by all-day singing records (Fig. 13). These data were taken concurrently with gonad data (Fig. 12) from March 1971 through January 1972. The main lekking season of *P. superciliosus* extended from December through July, dropping off sharply in

August; very low levels of lek activity occurred from September through November. The data thus include most of one lekking season, the period of low activity between seasons, and the start of the next season. Highest activity levels were recorded at the start of the lekking season, probably the period of territory establishment. During most of the season, singing activity fluctuated considerably, possibly on a day-to-day basis. Weather sometimes affected the amount of lek activity considerably; during midday lek activity was higher on cloudy than on sunny days, but lowest on rainy ones. In general, however, singing activity on the lek remained high through July.

The observed lekking season of *P. superciliosus* thus coincides quite closely with our estimates of the nesting season, allowing for possible year-to-year variation. The main difference between male and female reproductive seasons is that lek activity reached high levels practically as soon as the lekking season started, whereas nesting reached its peak one to two months later. Egg-laying and lekking declined more or less simultaneously in August.

The seasonal patterns of lek activity and nesting were strongly correlated with blooming seasons of important food plants (cf. Fig. 8). In particular, the start of the lekking season agreed closely with the onset of intense flowering by *Heliconia pogonantha*. Nesting peaked later as other important food plants such as *H. wagneriana* came into good bloom and the overall level of nectar availability increased considerably. This later peak is not unexpected as the energetic cost of nesting doubtless is greater than that of holding a lek territory. While early territorial establishment is advantageous to the male, early nesting may create severe energetic problems for the female, including depletion of her own energetic reserves to a dangerously low level (Ward 1969; Fogden 1972; Wolf and Wolf 1976). The major decline in lekking and egg-laying occurred when *H. pogonantha* was no longer in good bloom but other major food plants (e.g., H-16, *H. imbricata*, H-3, and *Costus malortieana*) were at their peak or still increasing. Females laying eggs at this time would still have another month of caring for nestlings and fledglings should the nest succeed. Young fledged later than September would be faced with low and still declining nectar resources and might well incur a reduced chance of survival.

The molting season was not clearly defined in *P. superciliosus*, and birds in molt could be found at any time of year. The peak of molt for the population was June and July but individuals differed by as much as 8 months in the timing of their 4-month molting seasons (in accordance with the long breeding season) (Stiles and Wolf 1974). Molting and breeding seasons of the population overlapped broadly, but the timing of these activities was largely independent among individuals. However, females sometimes interrupted molt to nest, while many males were active on the lek and molting simultaneously, another indication of the higher relative costs of nesting over lekking (Stiles and Wolf 1974).

Daily Pattern of Lek Activity

Levels of singing and territorial aggression on a *P. superciliosus* lek varied in a fairly regular and predictable manner during a day. Most activities were concentrated in three main singing sessions in the early morning, late morning, and

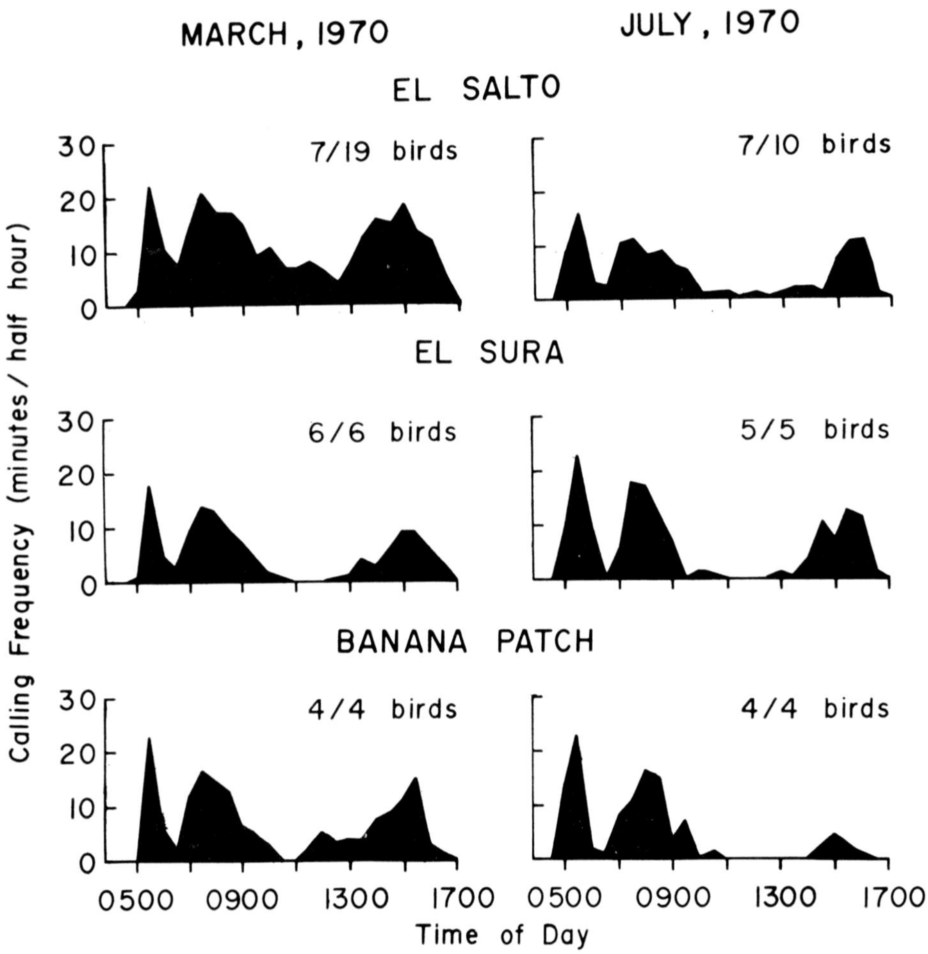

Fig. 14. Daily variation in calling activity on lek II (El Salto), lek Ia (El Surá), and lek Ib (Banana patch) for one day early and late in the 1970 lek season. Fractions indicate proportion of all birds on the lek included in the sample.

afternoon. This daily pattern was discernible over most of the lekking season, and for leks of different sizes (Fig. 14). Patterns of individual males often were less clear-cut. Singing and other activities of each resident were influenced more directly by songs and displays of neighboring residents and intruders. A resident's presence or absence at the lek was dictated largely by the length of his foraging flights away from the lek and, perhaps, day-to-day variations in the flowering of clumps included on his foraging route. Thus, the periods of most intense activity varied somewhat among different individuals on the same day (Fig. 15) or for the same individual among days (Fig. 16).

The early morning singing period began at dawn and usually lasted 30–50 min. Residents apparently did not roost on their territories, as in several nocturnal visits to leks we never found a roosting *P. superciliosus.* Rather, the residents flew from their roosts to the lek just as it began to get light. For a few minutes prior

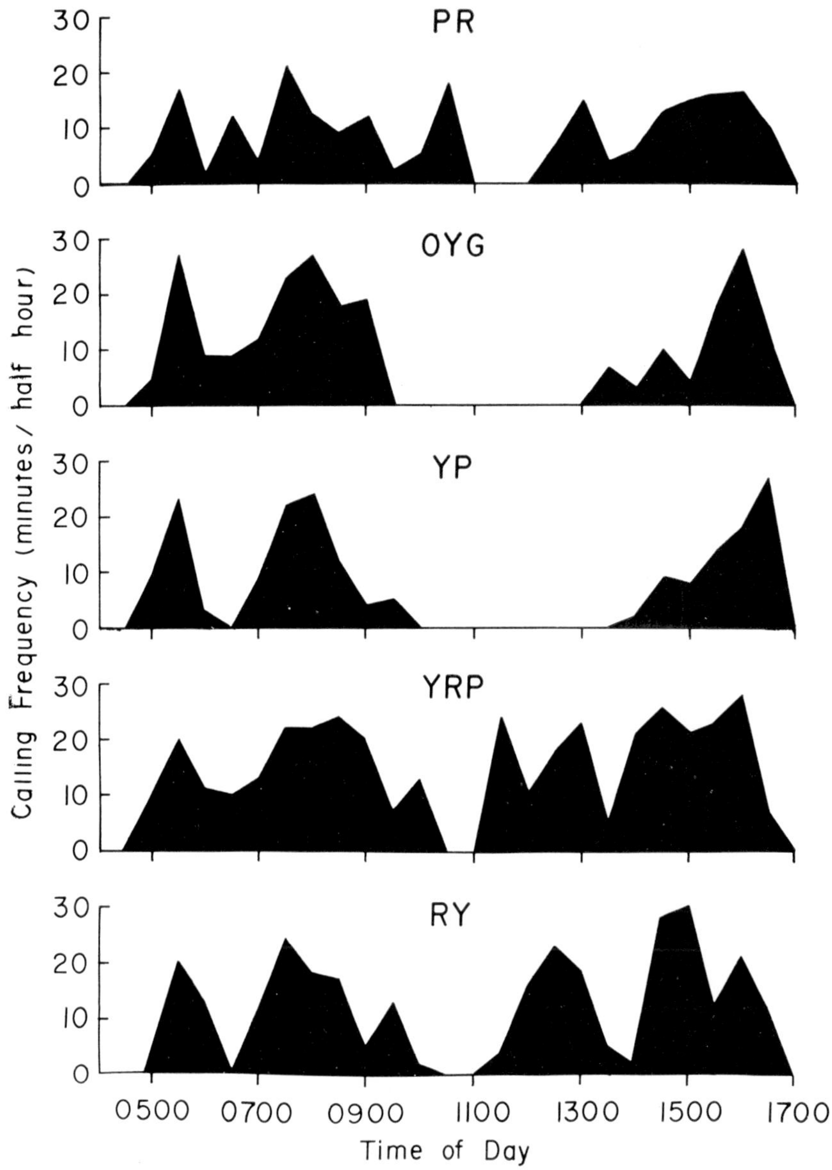

Fig. 15. Variation in daily calling activity patterns of five resident males on lek Ic on 29 March 1971. Locations of these males are shown in Figure 21.

to the onset of singing one could hear flight calls of hermits in the still-dark understory. Most of the residents used relatively high, exposed song perches at this time. Singing began at about 0515–0520 during the short days of December through January or February, and around 0505–0510 during the longer days of May through July. Cloudy, dark weather at dawn could delay the start of singing for several minutes.

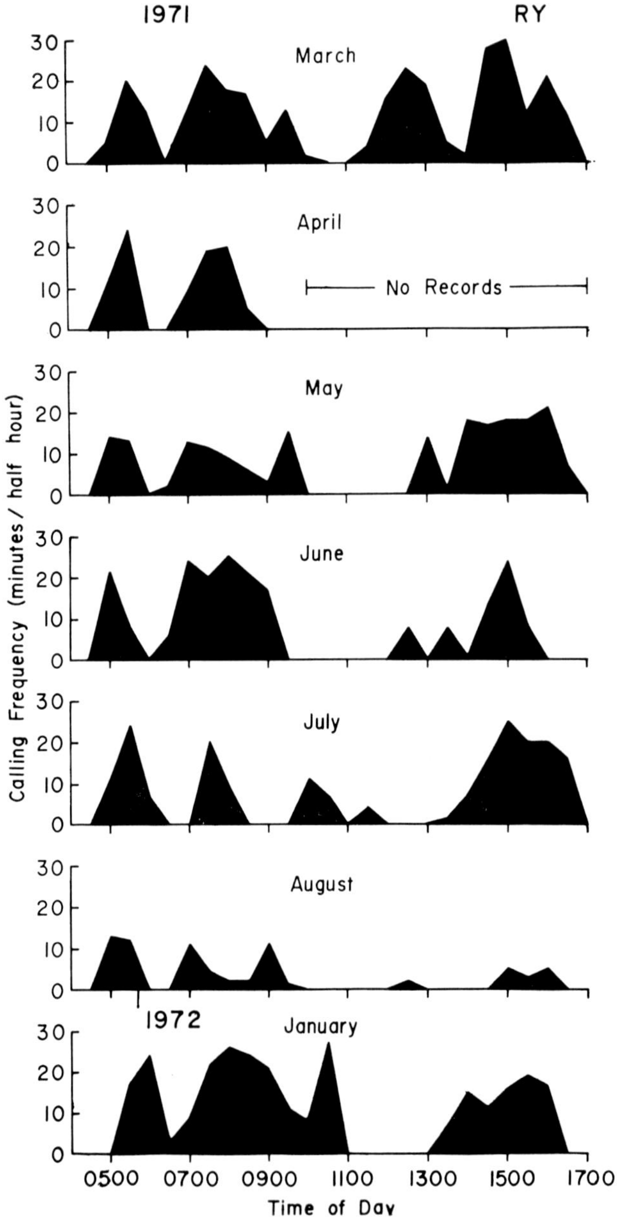

Fig. 16. Variation in daily calling pattern of male RY on lek Ic through the end of the 1971 lek season in August and the start of the 1972 season in January.

Usually one or two birds, not always the same individuals, initiated singing and after a few minutes, most or all lek residents joined in. The most intense and synchronous singing activity for the lek as a whole during the day occurred in the following 10–15 min (Figs. 14, 15). This dawn chorus declined as the light intensity increased to the point where an observer could distinguish the colors of

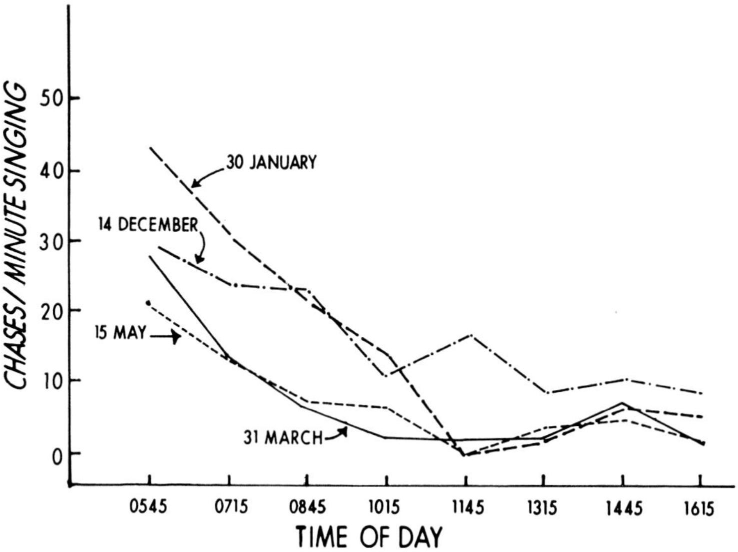

Fig. 17. Frequency of chase activity on areas observed during all-day watches at lek Ic. Chases are scaled by total minutes of singing for all birds observed; this corrects for absences of males and variations in numbers of residents observed on each date. The data indicate both a seasonal and daily decline in chases. Data are plotted at the midpoints of 90-min observation periods.

back paint and tags of the singing birds. It was now apparently light enough in the understory to permit the birds to engage in high-speed chases through the dense vegetation. A period of very intense aggressive activity followed in which territorial interactions and chases reached peak levels (Fig. 17). Singing activity continued at a high level, and the lek became a bedlam of songs, chatters, squeals, chase notes, and bill-pops. A given resident sang for a few minutes at most before going off to join a chase, challenge, or display at a neighbor, or an intruder. The mist-net samples (Table 10) indicate that virtually all of this early-morning activity was by resident males; females and nonresident males were only rarely captured at the lek during these hours. Probably this activity served to redefine the ownership of song perches and territories.

Most resident males almost simultaneously left the lek between 0545 and 0610 to forage, and for about 1 hour the lek was virtually deserted. Except for the few birds with suitable flowers on their territory, most residents probably had not fed before this, having been continuously present on the lek since dawn. We doubt that residents fed before arriving on the lek because of very low light intensities. Mist-netting and censuses in feeding areas away from leks revealed a burst of *P. superciliosus* activity beginning about 0600–0615.

As males returned to the lek following this "breakfast break," the synchrony of activity was reduced considerably. Throughout the rest of the day, each resident alternated periods of singing and interaction on the lek (usually 10–30 min) with absences of a few minutes to an hour or more. These absences, presumably for foraging, were longest and most frequent at midday and separated late morning and afternoon calling periods of most residents (Fig. 15).

TABLE 10

PHAETHORNIS SUPERCILIOSUS MIST-NETTED AT LEK I, 1971–1973

	Dec–Jan	Feb–May	Jun–July	Aug–Nov	Total
Adult males					
Resident territorial	29	114	33	2	178
Resident nonterritorial	0	0	6	14	20
Nonresident	10	7	1	2	20
Immature males					
Territorial	0	0	12	10	22
Nonterritorial	0	13	26	26	65
Adult females	4	12	3	4	23
Immature females	0	2	6	6	14
Total	43	148	87	64	342

The later morning singing period usually showed a well-defined peak about 0800–0830, but fairly high singing levels usually continued until 0900 or later (Figs. 14, 15). Chases are less frequent (Fig. 17), and the periods of uninterrupted singing averaged longer than in the early morning. Females apparently visited the lek most often during the late morning. Of the 19 females mist-netted on lek I during the lekking season (Table 10), 1 was caught between 0515 and 0700, 10 between 0700 and 0900, 6 between 0900 and 1100, and 2 between 1100 and 1200 (nets were usually taken down at noon). All observed copulations occurred between 0730 and 0930.

The onset and intensity of the afternoon singing period were quite variable, but the latter usually peaked around 1530–1600 (Figs. 15–17). Synchrony between males was further reduced, and active territorial defense, including chases, was rare (Fig. 17). Thus, a resident often engaged in 15–30 min of unmolested, continuous singing after arriving on the lek, and the peak levels of song for the lek as a whole could be comparable to those of the morning. Residents also spent considerable time sitting quietly on their song perches in the afternoon. Few females visited the lek at this time, but later in the breeding season young males were especially conspicuous on the lek in early to midafternoon.

Activity ceased gradually in the late afternoon. One by one, the residents left their territories and flew from the lek. By 1700 usually only one or two birds were still present, and all singing had stopped by approximately 1715. A resident might remain on his song perch for several minutes after he stopped singing. At the time lek activity ceased for the day, it was considerably lighter than when singing started at dawn.

The daily course of lek activity by *P. superciliosus* showed certain definite relationships with the nectar secretion patterns of major food plants (cf. Fig. 9). In particular the ratio of active territorial defense (numbers of chases) to territorial advertisement (bird-minutes of singing) declined through the day from a peak in the early morning, as did the nectar production curves (Fig. 17). The most energy was expended in territorial activity just prior to the period in which

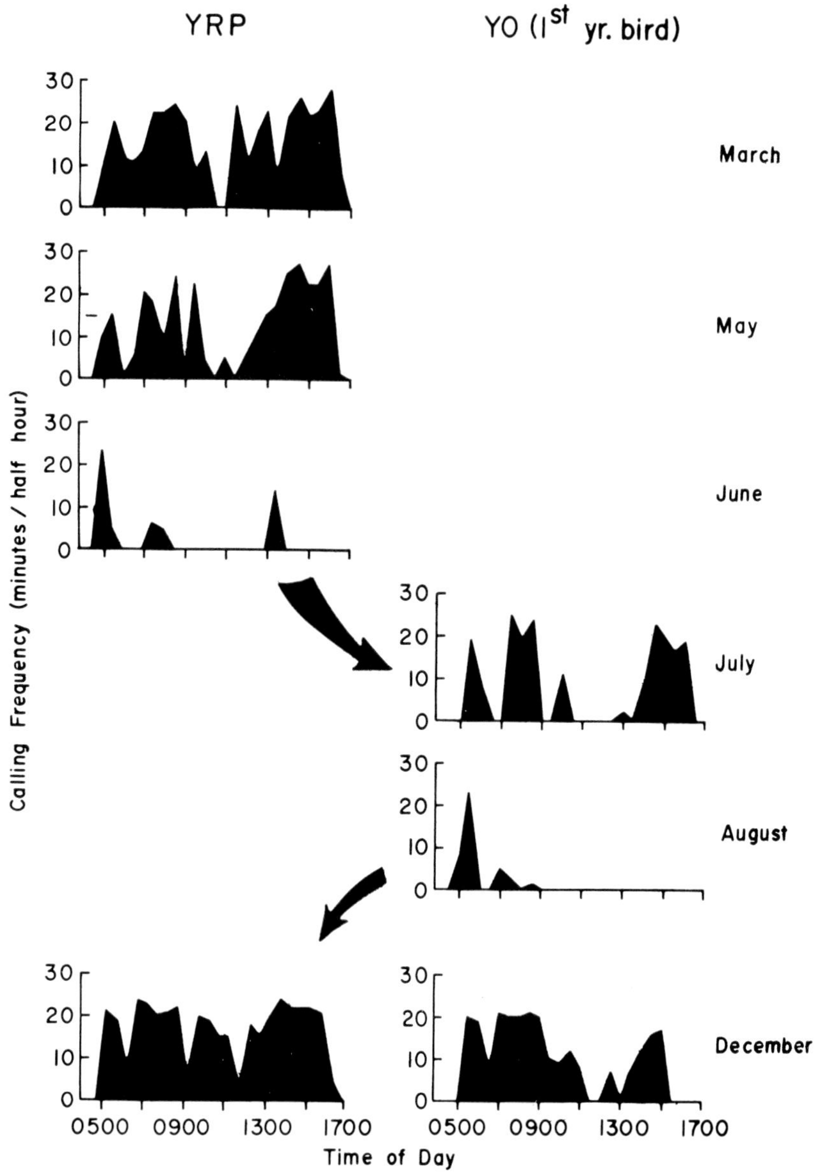

Fig. 18. Seasonal variation in daily calling activity from one territory on lek Ic. When YRP left the lek in mid-July it was quickly replaced by YO, a young bird with a relatively short period of high activity at the end of the season. The territory was reoccupied by YRP in December 1971 as the season started and YO, now marked as YRO, held an adjacent territory (see Figure 22).

nectar supplies were at their daily maximum, the "breakfast break." Conversely, in the later afternoon when little nectar was available, little energy was expended in chases as most birds engaged in long bouts of energetically economical singing. In addition, flowers of some *Heliconia* species were still too tightly closed at dawn

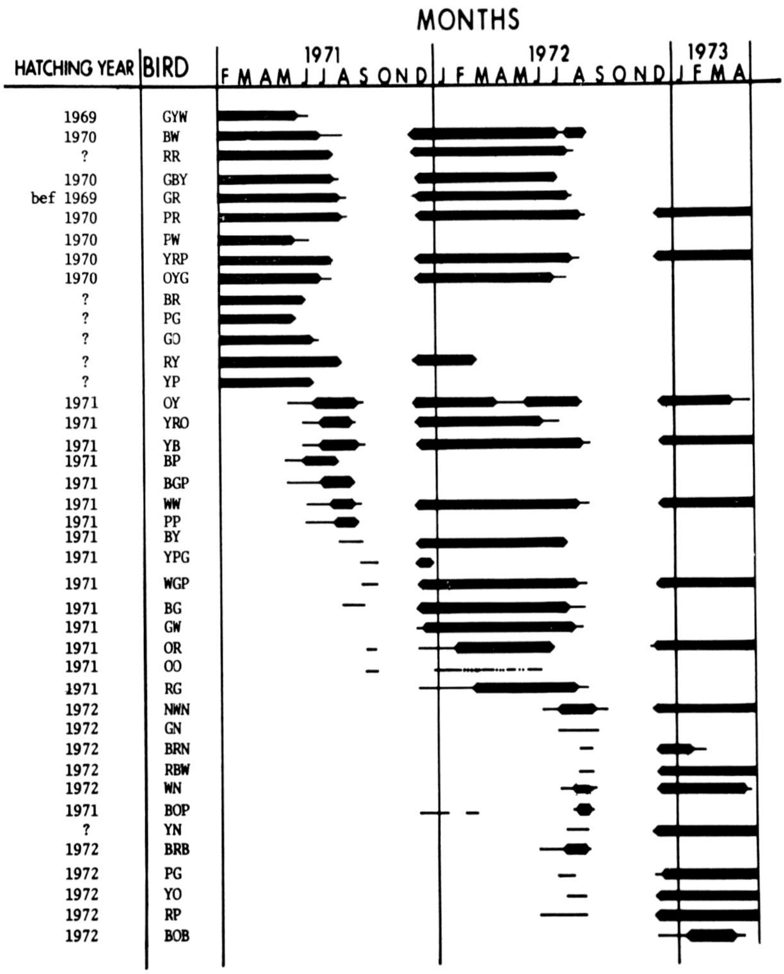

Fig. 19. Presence and territorial occupancy of male *P. superciliosus* on lek I, February 1971 to April 1973. A thick bar indicates territorial occupancy, a line indicates the male was present but not controlling a territory.

to permit easy access to the nectar by hummingbirds. *H. imbricata* flowers, for instance, did not open fully until 0545–0600, about the start of the "breakfast break."

The daily pattern of lek activity also varied seasonally. At the peak of lek activity (December and January), singing was practically continuous through the day, and the midday break was reduced or absent (Fig. 18). Through most of the lekking season (February through May or early June), the three singing periods were quite distinct. As the end of the lekking season approached, the residents progressively restricted their lek activity to the early morning (e.g., YRP in Fig. 18). Just before leaving the lek for the season, the residents appeared and sang briefly only in the early morning; for the rest of the day young males hatched earlier in the breeding season contributed most to the activity of the lek.

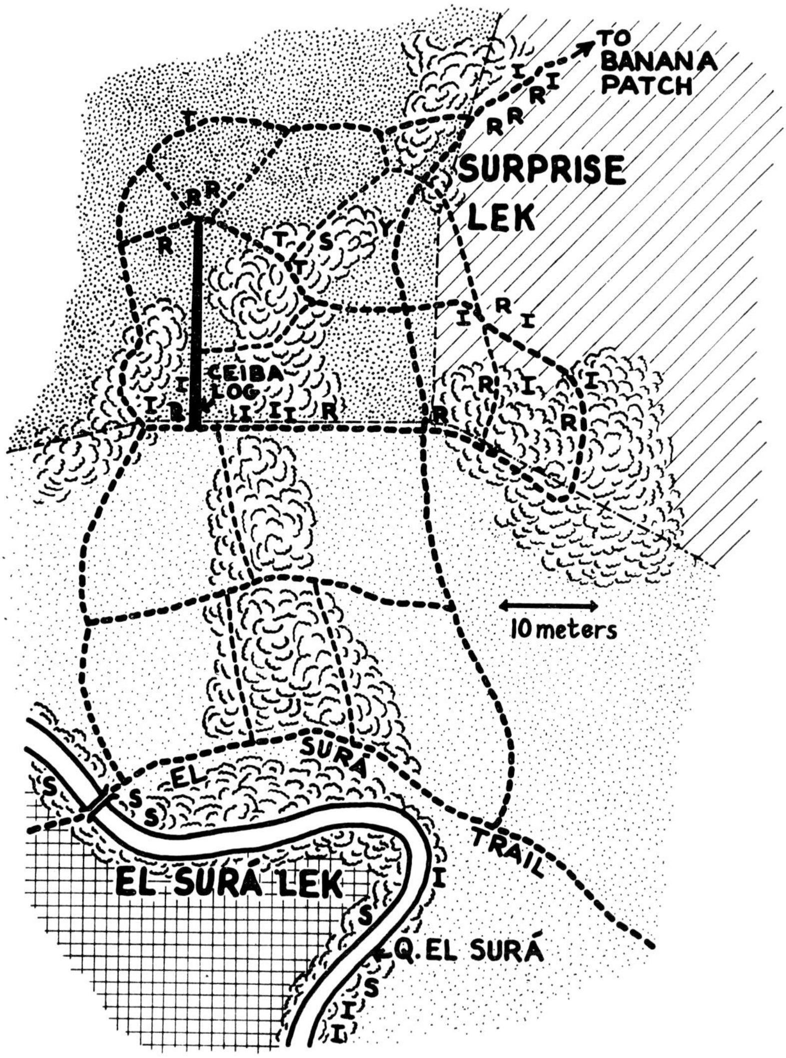

Fig. 20. Locations of lek Ia, "El Surá," and lek Ic, "Surprise," showing vegetation types, major landmarks, and distribution of major food plants. Heavy stipple = forest; light stipple = cacao; cross-hatching = arboretum (open grassy understory, trees left standing); wavy lines = dense thickets. Letters denote clumps of *Heliconia* plants as follows: I = *H. imbricata*; R = *H. pogonantha*; S = H-3; T = H-18.

LEK DYNAMICS

Seasonal Variation in Lek Composition

Between February 1971 and April 1973, we tried to keep all the residents of lek I marked, and to mark new individuals as they appeared, by setting mist-nets in the lek once every 1–2 months (Fig. 19). Both before and after mist-netting, we ascertained the territorial status of as many individuals as possible. Because we captured most or all of the lek residents each month, we doubt that birds

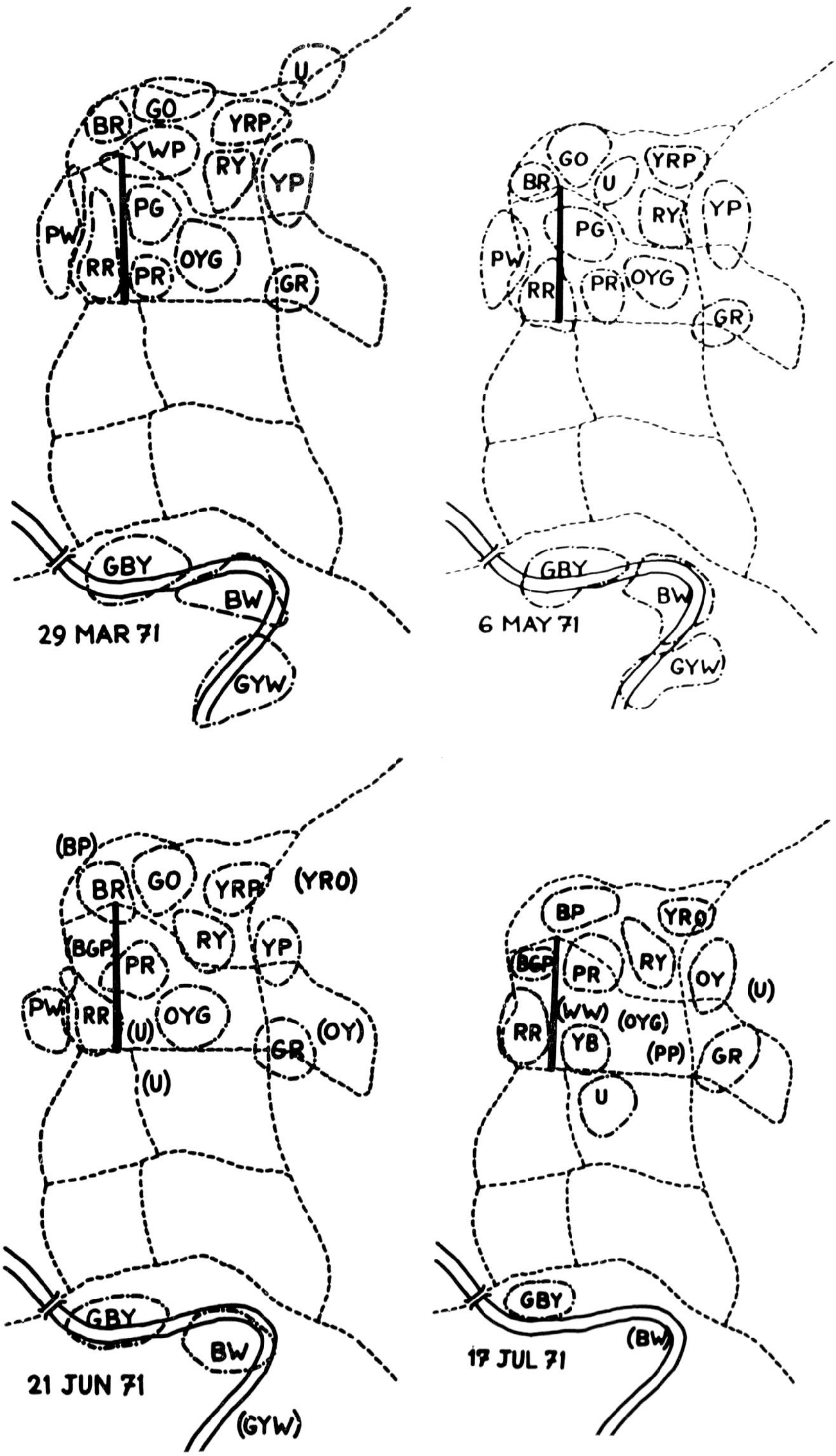
U
GO
BR
YRP
YWP
RY
YP
PG
PW
OYG
RR
PR
GR
GBY
BW
GYW
29 MAR 71
GO
YRP
BR
U
RY
YP
PG
PW
PR
OYG
RR
GR
GBY
BW
GYW
6 MAY 71
(BP)
BR
GO
YRP
(YRO)
BGP
RY
YP
PR
PW
RR
OYG
(U)
GR
(OY)
(U)
GBY
BW
21 JUN 71
(GYW)
BP
YRO
BGP
PR
RY
OY
(U)
RR
(WW)
(OYG)
YB
(PP)
GR
U
GBY
(BW)
17 JUL 71

learned to avoid the nets. Residents and intruders flew at the same heights, so we probably captured most or all of the intruders on a given day, as well. In view of the difficulty of identifying and sexing intruders visually, the mist-net data (Table 10) probably provide the most complete and unbiased possible sample of the individuals present on lek I on those particular days. Data from lek II (1970–72) and lek III (starting 1973) are consistent with the information derived from lek I (Table 10). We therefore believe that the lek I data are representative of the seasonal dynamics of *P. superciliosus* leks at La Selva.

The largest numbers of adult males were present on the lek at the start of each lekking season (all young of the previous breeding season could now be considered adult). In December and January, 70–80% of these males obtained territories, although some failed to hold them (e.g., YPG in 1972, BRN in 1973; see Figs. 20–24). By February relatively few nonterritorial males were about, mostly persistent intruders that never established themselves on the lek (e.g., OO in 1972). Where the unsuccessful males went is unknown, although a few became residents of another lek (e.g., WY, marked on lek I in February 1971, and a resident of lek II by April 1971). Once territories were defined and nonresident males excluded, the tempo of lek activity stabilized at a slightly lower level until May or early June (Fig. 13). Changes in lek composition during most of this period were due to occasional deaths of old residents and additions of new residents at the periphery of the lek. These peripheral territories often were unstable in space and time, and the birds establishing them were individuals that appeared on the lek late in the season as well as those that failed to obtain territories earlier (e.g., unmarked resident(s) in March and May 1971, RG in 1972; Figs. 21, 22).

A major change in lek composition occurred with the appearance of young males hatched earlier in the same breeding season. A few appeared on the leks as early as March (possibly 2–3 months postfledging); their numbers began to increase in mid- to late May (Table 10). When these birds first appeared, their songs were variable, unstructured, and of no definite song type. They often sang from various points on the lek, partly as a result of recurring displacement by residents, especially during periods of low lek activity, as at midday. In periods of more intense activity, young birds sang from high perches, often 6–10 m above ground, where adult males seldom sat except early in the morning. After a few weeks on the lek, some young males established territories on the lek periphery. However, relatively few young males became territorial before mid- to late June.

The reduction in activity of an adult male prior to departure from the lek probably reflected his own physiological condition rather than conditions on the lek as a whole. In late June some adults were still fully active, while others had reduced activity to varying degrees (Fig. 19). A male that left the lek between

---

←

Fig. 21. Locations of territories of males on leks Ia and Ic in March, May, June, July, 1971. The dashed lines represent the territory of each male (= the area containing all his song perches) on the given date. Parentheses indicate a male that sang or interacted briefly at a particular site but was not consistently territorial on that date. Specific identity, if known, of each male is indicated by his color code; unmarked males are indicated by a U. Note the tendency of territories to occur in dense vegetation and not to be centered around food plants.

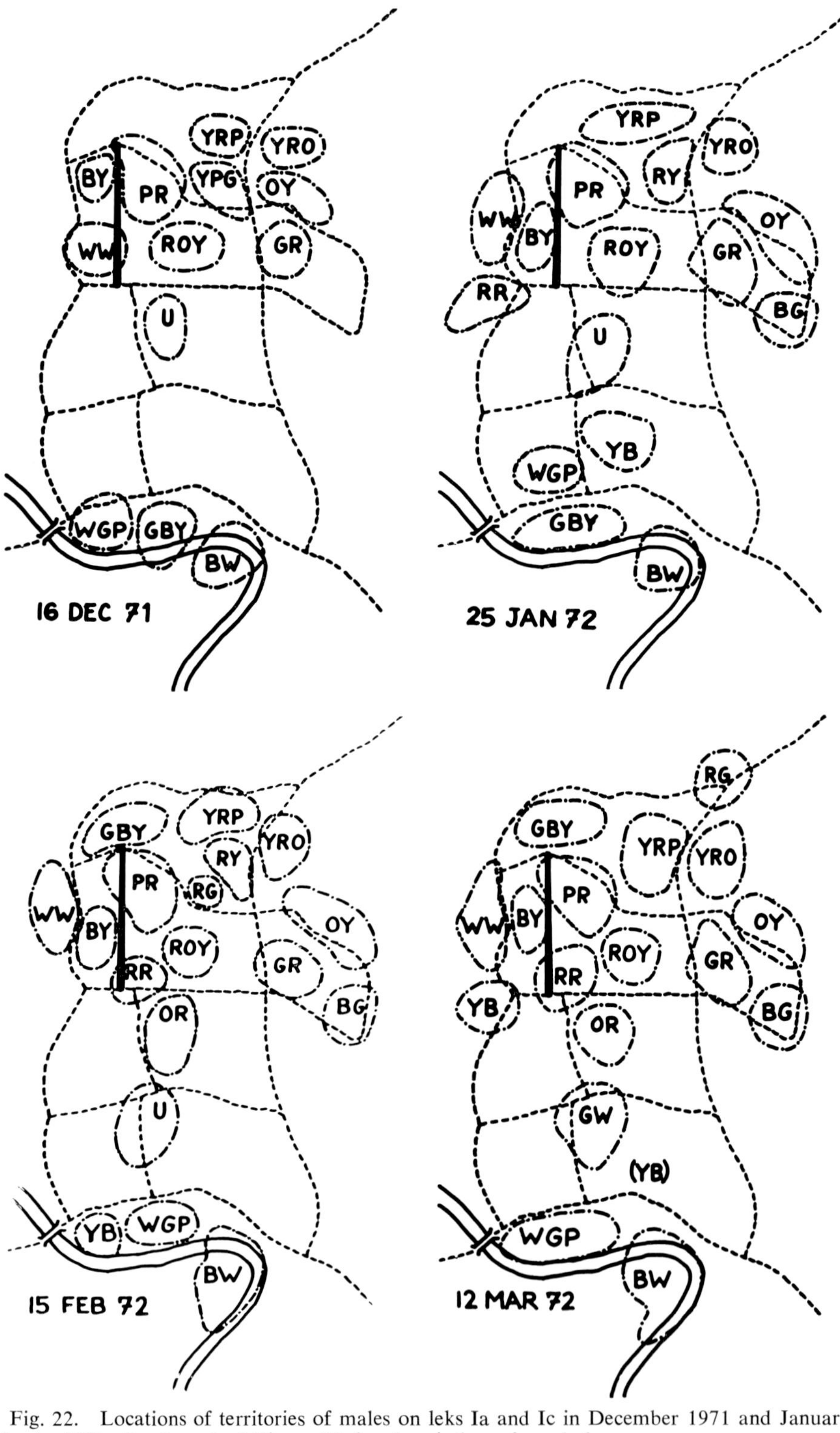

Fig. 22. Locations of territories of males on leks Ia and Ic in December 1971 and January–March, 1972. See legend of Figure 21 for description of symbols.

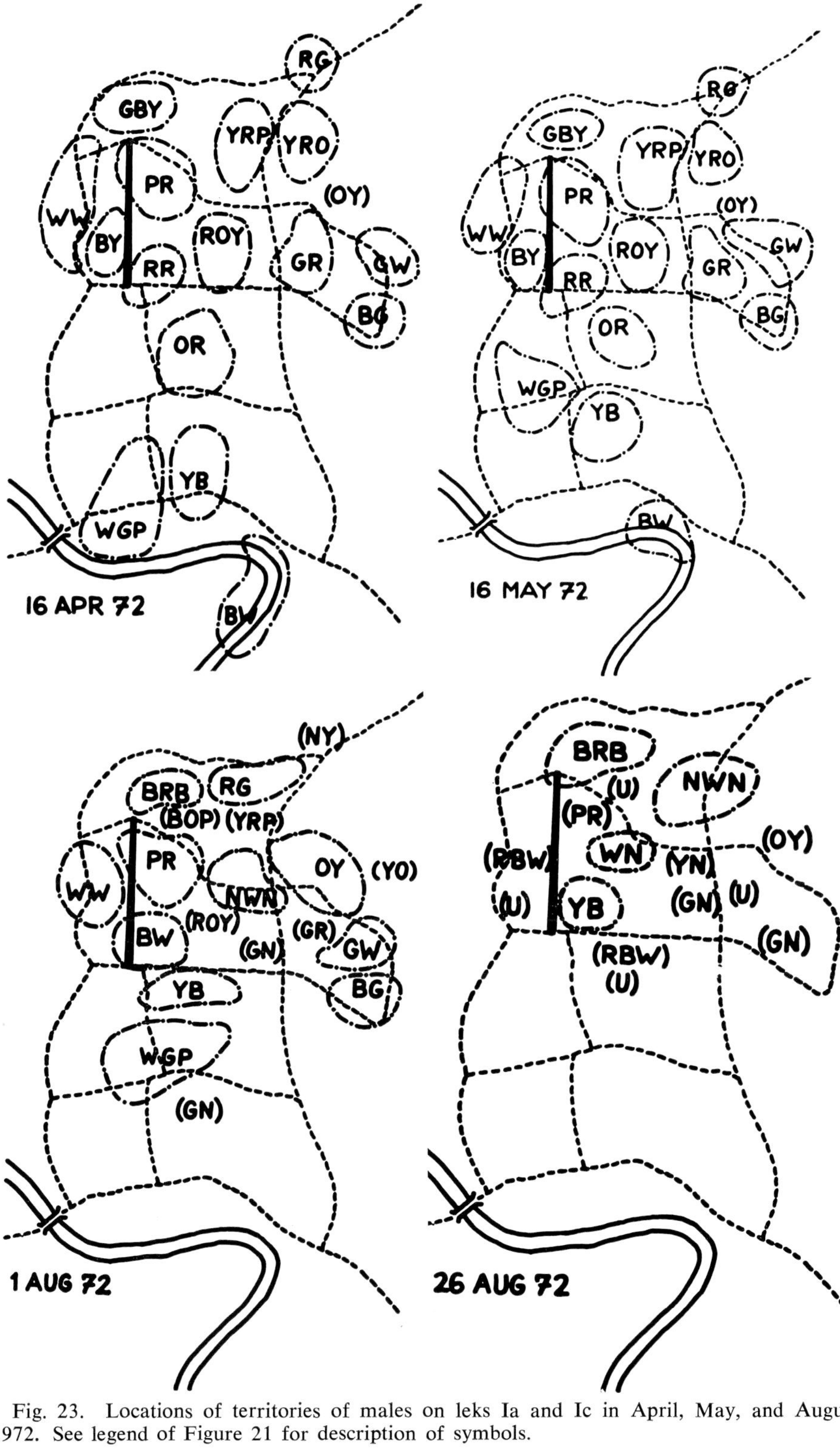

Fig. 23. Locations of territories of males on leks Ia and Ic in April, May, and August, 1972. See legend of Figure 21 for description of symbols.

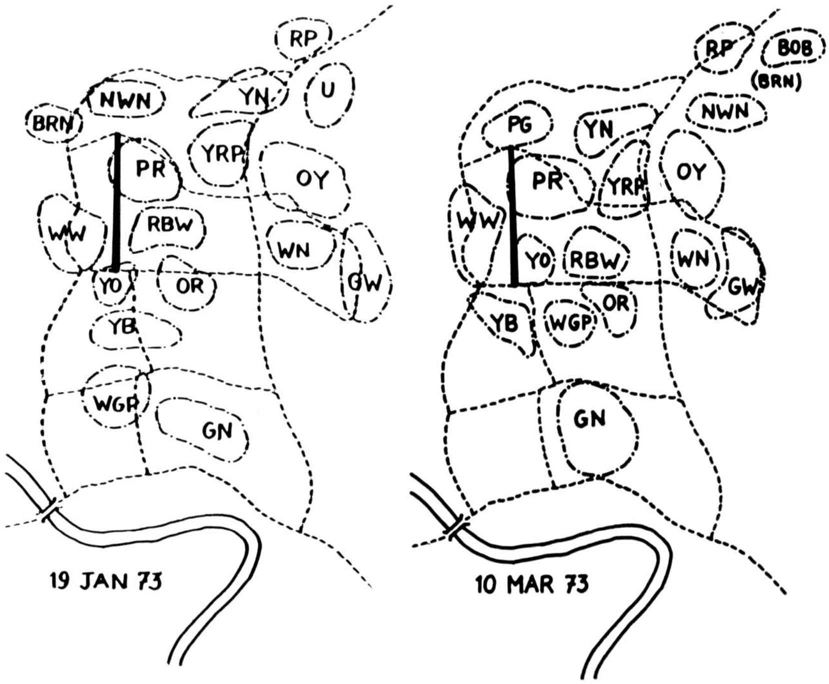

Fig. 24. Locations of territories of males on leks Ia and Ic in January and March, 1973. See legend of Figure 21 for description of symbols.

1 June and 1 July was less likely to return the next season than a male leaving after 1 July (Fisher's exact probability test; $p = 0.022$). The departure of adults sometimes coincided with the start of molt or the period of heavy body molt, but some males showed no relation between the intensities of lekking and molt (cf. Fig. 3 of Stiles and Wolf 1974). The vacated areas at the end of the lekking season often were taken over by young males, and a sharp increase in the number of young males holding territories occurred at this time (Table 10).

Through July and early August lek composition changed rapidly so that by August most territories were held by birds of the year. When an adult left his territory and a young male moved in, the latter usually sang at a much higher rate than the adult, at least initially (Fig. 18). A brief resurgence of song could occur on the lek as a whole as adults were replaced by young birds (note the small peak in July, Fig. 13).

By late August lek activity had declined precipitously. All adults either had left the lek or appeared infrequently. The immature males also were leaving the lek, and only a few still sang regularly enough from a localized area to be considered territorial. In September singing was brief and irregular, and no birds were territorial. Males, usually immature, entered the lek for a few minutes, sang briefly from several territories of the previous year, and left. Only in the early morning was more than one bird regularly present. In October and November, the lek was virtually deserted. One heard at most a song or two at dawn, then for the rest of the day only flight calls and, rarely, a chase, presumably involving birds whose foraging routes happened to pass through or close by the lek area. Activity on the lek at this time probably was similar to that in any other comparable habitat at La Selva.

Male *P. superciliosus* returned to the lek within a week or two at the start of the next lekking season (late November or December in most years; Fig. 19). On 1 December 1971 lek I was practically deserted, although a brief song was heard along the Quebrada El Surá in midmorning. On 13–14 December 13 birds were vigorously territorial, and 7 intruders were identified, 4 of which eventually became residents.

Females apparently visited the leks relatively infrequently, as we suspected from the scarcity of observed copulations. From Table 10 we calculate that an average of 1.5 adult females per day appeared on the lek during the lekking season, compared with 0.55 per day when the leks were not active. Thus, unless females not ready to mate avoided the lek during the lekking season, only about one female per day visited the lek to mate. If a lek of 15 males serves 10–30 females, each female must come to the lek on the average of once every 2–4 weeks. This seems a reasonable estimate because a successful nesting attempt of *P. superciliosus* requires about 2 months, but at least 75% of the nests of this species fail and thus require less time (Skutch 1964a). The length of time expended in an average nesting attempt thus is probably about three weeks, as females probably renest fairly soon after a failure (D. Snow and B. Snow 1964, 1973; Stiles 1973).

Survivorship and Recruitment of Lek Residents

We examine two aspects of survivorship in *P. superciliosus*: (1) the turnover of lek males between years, and (2) the seasonal pattern of mortality of lek residents including immature birds that establish territories in their first season. Females and intruders were impossible to study in this regard because they were not localized and were recaptured only rarely and inconsistently. Determining seasonal mortality patterns is made more difficult because many males disappeared completely for 3–5 months between lekking seasons, precluding regular censuses. Therefore, one can only compare survivorship over the lekking season with that between lekking seasons. Measuring survivorship over the nonlekking season assumes that surviving residents return to the same lek at the start of a lekking season. We have no evidence of established residents switching leks nor have we ever encountered a male that was territorial one year and nonterritorial the next.

Our most reliable survivorship data are from lek I between 1971 and 1973 (Table 2). Survivorship was significantly higher through the lekking season (ca. 90%) than it was between lekking seasons (ca. 55%) ($\chi^2 = 8.69$; $p < 0.005$). It is unlikely that human interference produced this difference because such interference, in the form of monthly recapturing and handling, was much greater during the lekking season. Also, no bird so completely lost its markings during the nonlekking season that it was unidentifiable when it returned to the lek in December. We have recapture data for a number of individuals during the nonlekking season. Every male that failed to return to the lek in December apparently disappeared in the latter part of the nonlekking season. The data suggest that up to 75% of

TABLE 11

SURVIVORSHIP OF RESIDENT MALE *PHAETHORNIS SUPERCILIOSUS* WITHIN AND BETWEEN LEKKING SEASONS: LEK I, 1971–72

	Adult males		Immature males	
Date	Surviving	Died during interval	Surviving	Died during interval
Feb 71–Aug 71	12	1	–	–
Aug 71–Dec 71	7	5	4	3
Dec 71–Jul 72	16	2	–	–
Jul 72–Dec 72	8	8	5	3

the annual adult mortality occurred in October and November, the time of year when flowers were scarcest (Fig. 8); food shortage may play a major role in limiting the *P. superciliosus* population at La Selva. Young males survived as well as older males over the nonlekking season. This is much less surprising in *P. superciliosus* than it would be in a species where adults hold food-containing territories during the nonbreeding season. Given the seasonal and year-to-year shifts in flower abundance to which *P. superciliosus* were exposed (Stiles 1975, 1978a), old and young males faced the season of food shortage on a far more equal footing than is often the case in territorial species. A potential bias may occur in that we consider only young males that established lek territories; these may have been the strongest and most vigorous young birds, possibly with above-average survivorship for their age-group. In any case it seems evident that if a period of high postfledging mortality existed, it came prior to the time when young males established themselves on the lek. Combining the between-season and within-season survivorship percentages, one finds an annual turnover rate of males on a *P. superciliosus* lek of about 50% (Table 11).

If survivorship after the postfledging period is independent of age and half the males on a lek are replaced each year, then a male *P. superciliosus* must live, on the average, for only one to two seasons. In any one lekking season half of the residents will be in their first full season on the lek; very few birds will be over 3 years old (Table 12). Considerable recruitment must occur each year to maintain the leks at a relatively constant size (Table 2).

TABLE 12

AGE STRUCTURE OF *PHAETHORNIS SUPERCILIOSUS* ON LEK I IN 1971–1973 LEKKING SEASONS

No. residents aged	1971	1972	1973
1 yr	6	8	8
2 yr	1	6	5
3 yr	1	–	2
4 yr	–	1	–
Unknown (probably older than 1 yr)	5	1	1
Total	13	16	16

Our best data on recruitment again come from lek I between 1970 and 1973, when we usually were able to mark young males as they appeared on the lek and to follow their subsequent attempts to establish residency. We have some knowledge of the histories of 32 residents of lek I prior to the time they established residency. Only YN was an adult when he first appeared on the lek; all others were marked as immatures and became lek residents in their first full lekking seasons. The lek experience gained by a young male during his first season may influence his chances of subsequently becoming a resident, as is demonstrated by the histories of the 40 young males that appeared on lek I during the 1971 and 1972 lekking seasons (Table 13). All surviving young males who held territories as immatures, and most young males who sang or interacted, became residents on lek I in the next season. Of the 12 young males without known experience on lek I, 5 became residents of leks III or IV and may have had lek experience there; at least two of these were territorial on lek III. Only 3 of the 12 became residents of lek I (Table 13).

The timing of a young male's first appearance on the lek can affect his chances of eventually becoming a resident by influencing (1) his survivorship, and (2) the amount of lek experience he gains as an immature. Relatively more young males arriving on lek I in March–May obtained territories as immatures than did those appearing in June–August, or in September or later (3/7 = 43%, 7/26 = 27%, and 0/7 = 0%, respectively). However, survivorship to the next breeding season showed the opposite trend in the three groups (57%, 65%, and 71%, respectively: see Table 13). Perhaps the early birds had less energy reserved going into the lean season because of their longer periods of strenuous lek activity. The overall result was that recruitment of lek residents was greatest from the June–August group, both relatively and absolutely (14/26 = 54% vs. 3/7 = 43% for both early and late groups). Birds that appeared on the leks late could gain territories the next year largely because many of their older and more experienced fellow yearlings failed to survive. Any young male surviving the lean season had a good chance (20/26 = 77%; Table 13) to become a lek resident, and the few that failed to do so were among the less experienced and presumably younger members of their year-class. Unfortunately, we have little information on the survivorship of nonresident males, but if a young male failed to become a lek resident in his first full lekking season, his chances of doing so in a subsequent season apparently were low. Practically all recruitment of new lek residents was from the ranks of yearling males (Table 11).

Intermale Relations on the Lek

Possible indicators of the relative status of lek residents include morphological or other attributes of the males, variations in their territorial behavior, or differences in the quality of their territories. Obtaining information on these points is difficult. Differences between males are likely to be subtle, and the dense vegetation of the lek makes it impossible to observe several territories simultaneously or to follow interactions for any distance. In other species of lek birds dominant males do most of the mating on the lek (Hogan-Warburg 1966; Lill 1966; Robel 1966; Wiley 1973). However, we were unable to evaluate mating success because we so seldom saw copulation.

TABLE 13

SUBSEQUENT STATUS OF YOUNG MALE *PHAETHORNIS SUPERCILIOSUS* FIRST APPEARING ON LEK I IN 1971 AND 1972

Experience on lek I in first season	Became resident on: Lek I	Lek III or IV	Never became resident on any lek	Died
Held territory	6	0	0	4
Sang and interacted but not territorial	5	1	2	4
Neither sang nor interacted	3	5	4	6
First appeared on lek				
March–May	2	1	1	3
June–August	10	4	3	9
September and later	2	1	2	2
Totals	14	6	6	14

Several lines of indirect evidence suggest that males in central territories on a *P. superciliosus* lek tended to be dominant to males with peripheral territories. If a vacancy occurred in the center of a lek during the main lekking season, the space was always filled by the shifting and expansion of the territories of neighboring residents. Peripheral males might attempt to establish a territory in the center should a central resident die (e.g., RG when RY disappeared in March 1972; Fig. 22), but they were never successful in our experience. By contrast, vacant peripheral territories were filled, if at all, by formerly nonterritorial birds, and newly resident males established themselves on the periphery. Also while central territories had relatively stable locations and boundaries, peripheral territories often were very unstable in space and time (e.g., YB in the 1972 lekking season; Figs. 22, 23).

When young males began to settle on the lek in May and June, they always established territories on the lek periphery, at least until the central residents began leaving their territories at the end of the lekking season. We never saw a young male chase a resident adult and there was never any suggestion of several young males moving into a territory and wearing down the defenses of an adult, as apparently happens in some North American hummingbird species (Stiles 1973). A young male taking over the territory of a resident was also setting the stage for a contest with that resident for possession of the territory should both return for the next lekking season. In 20 of 24 such cases (Table 14) the old resident regained his territory at the outset; in a few cases the resident moved to a more central territory when another resident failed to return. The one exception occurred in 1972, when RR apparently returned late to the lek and WW and BY became established in most of his former territory. Over the next month RR forced WW and BY back and eventually controlled practically the same area he had in 1971 (Fig. 22).

Young males establishing residency on the lek at the start of their first full lekking season were much less likely to regain the same territory they had held at the end of the previous season (Table 14) because they were almost invariably

TABLE 14

SHIFTS IN LEK TERRITORIES OF RESIDENT MALE *PHAETHORNIS SUPERCILIOSUS* IN RELATION TO PREVIOUS STATUS[1]

	Status of male in previous year		
	Adult resident	Young territorial	Young nonterritorial
At start of lekking season in second year:			
Regains old territory	20	5	1
Takes more central site	3	1	1
Takes more peripheral site	1	7	3
During lekking season in second year:			
Shifts toward center	5	5	2
Shifts toward periphery	1	1	0

[1] Data are with respect to site of most frequent singing and interacting in previous year.

displaced if the old resident returned. If a returning resident did not reoccupy his old territory at the start of a lekking season, he was more likely to move to a more central spot (3 of 4 instances) while a young male was more likely to occupy a more peripheral site (7 of 8 instances). Later shifts of both old and young birds (following the disappearance of established residents) were nearly always centripetal in direction. Eight of 18 young males that sang or interacted fairly persistently at a site but were never really territorial there obtained the same spots they had held as immatures or more central ones, while 23 of 24 returning residents did so, a highly significant difference ($\chi^2 = 10.73$, $p < .01$).

However, this dominance of old residents over young ones was not absolute and related mainly to contests for specific territories, where the incumbent resident might be expected to have the advantage. In a number of instances, a resident reoccupied his old territory instead of a more central site (presumably available because another resident failed to return), and a young male succeeded to the ownership of the central site. Usually the young male had occupied the site at the close of the previous lekking season (e.g., RBW, Figs. 23, 24). A high degree of site attachment on the part of residents is indicated, and the amount of previous experience on a territory (whether as a full resident or a yearling) may be important in determining a male's capacity to regain that territory in succeeding years. The chief means for residents to improve their position on the lek (i.e., moving to a more central location) is apparently to shift their territories as more central residents disappear during the lek season, rather than taking up presumably better territories at the start of the season.

Some individuals seemed more or less dominant or subordinate from the start. At the end of the lekking season the last residents to leave sometimes shifted to more central territories before finally departing. YB, as a young male in August 1971, moved into an area formerly controlled by OYG and RR. The following year YB was excluded from this area as the old residents returned; he spent the 1972 season as a peripheral resident between leks Ia and Ic (Figs. 21–23). At the end of the 1972 season YB again moved into this area following the departure of RR and BW (Fig. 23). In 1973 YB was again on the periphery of lek Ic after losing a contest with RBW for his apparently preferred spot. Other birds that remained more or less peripheral for two full seasons were WW and WGP, even though young males like YO and RBW were able to occupy more central positions. It seems that an individual's dominance status often may be established during his first year on the lek and may not change markedly thereafter.

There is considerable evidence that male *P. superciliosus* preferred to hold territories in areas where activity levels were high, generally the more central locations (see discussion). For instance, GR, the lone survivor of lek Ib in February 1971, deserted the territory he had held since 1969 and moved to a peripheral spot nearby on lek Ic. Lek Ia gradually passed from a major center of activity to a minor one through 1971 and 1972. With the exception of BW and GYW, males on Ia tended to be those unable to hold a position on Ic. GBY was a resident on Ic in February 1971, but soon thereafter lost his territory to RR. GBY then took over RR's territory on Ia, but moved back to Ic the next year when a peripheral resident, GO, failed to return for the 1972 lekking season. During the 1972 season, two other residents of Ia, YB and WGP, shifted their territories towards the periphery of Ic; by May BW was essentially alone on Ia. Finally in June 1972 BW moved to Ic and took over RR's territory, RR presumably having left for the season (Figs. 22, 23). Unfortunately, neither BW nor RR returned for the 1973 season. Lek Ia has remained deserted since BW left it.

As we were unable to measure dominance of a male directly, we have assumed from the preceding information for these hummingbirds and also from information on other lek species (e.g., Hogan-Warburg 1966; Wiley 1973) that spatial position on the lek often is a reflection of the male's relative dominance position. In general, dominant males tend to be more centrally located than subordinates. For the hummingbirds, position of a territory was defined by its spatial location relative to adjacent territories. A central territory had common boundaries with at least four other territories and at least 270° of its circumference adjoining other territories. An intermediate territory had common borders with two to four other territories. A peripheral territory had common boundaries with only 1–3 territories and 180° or less of its circumference adjoining other territories. For most relationships examined the spatial position of a male was taken as the most central territory held for two or more months in the first four months of the lekking season. However, for the relation between time spent singing and position while singing, we used the position occupied when the singing data were recorded.

We found no quality of the males that consistently correlated with position. Bill length is correlated with mating success in a sandpiper sexually dimorphic in this character but not in plumage (Jehl 1970), but there was no significant tendency for longer-billed male *P. superciliosus* to be more central (Table 15). Also no significant correlation exists between body weight and position (Table 15), in accord with the data on body weight and dominance from various passerine species (Smith 1976). Yet the nonterritorial category does contain a relatively higher proportion of small-billed, lightweight birds than the territorial categories. No color character in *P. superciliosus* is particularly variable except back color, which should play no role in displays or dominance interactions because displaying birds face one another (Fig. 7). The quality of loudness of a male's song may

TABLE 15

MORPHOLOGICAL AND BEHAVIORAL PARAMETERS[1] OF MALE *PHAETHORNIS SUPERCILIOSUS* IN RELATION TO LEK POSITION

	Position			
	Central	Intermediate	Peripheral	Nonterr. or failed to hold territories
Morphological parameters[2] (data from leks Ia, Ic)				
Sample size	10	16	10	8
Body weight (g)	6.09 ± 0.27 (5.7–6.4)	6.14 ± 0.25 (5.6–6.5)	6.08 ± 0.14 (5.9–6.4)	5.99 ± 0.25 (5.7–6.3)
Bill length (mm)	37.50 ± 1.00 (36–39)	37.50 ± 1.35 (35–39.5)	37.55 ± 0.89 (36–39)	36.69 ± 1.43 (35–39)
Behavioral parameters (data from leks Ib, Ic, II)				
Sample size	12	12	6[3]	
Minutes singing 0730–0930	76.83 ± 11.04 (49–87)	67.50 ± 16.02 (32–93)	41.67 ± 13.49[4] (23–58)	
Length of early morning foraging period (min)[2]	58.92 ± 13.07 (32–82)	59.58 ± 15.33 (30–82)	69.93 ± 11.14 (38–93)	

[1] Data given are mean, standard deviation, and range in each case.
[2] No means are significantly different by Student's *t*.
[3] Includes nonterritorial or failed to hold territories category.
[4] Highly significantly different from other means in group by Student's *t*.

be correlated with his dominance, and certainly intruders and young males sing differently from residents. However, among established residents the situation was complicated by dialect differences, and several central residents had rather soft songs.

Variations in territorial behavior included rate and especially persistence of singing, and differences in the frequency with which a male invaded the territories of his neighbors in relation to how often he had to defend his own territory against their invasions. If dominance on *P. superciliosus* leks is translated, directly or indirectly, into mating success, then it is important to compare and evaluate territorial behavior in that interval when females are most likely to appear on the lek, about 0730–0930. Data from all-day singing records indicate that peripheral males sang significantly less persistently during this period than did more central males (Table 15). Peripheral males tended to have shorter song bouts as well, at least in part because they often were challenged by more central males as soon as they began to sing. Central males sang for longer periods uninterrupted by invasions of other residents. On several occasions we noted that prior to losing all or part of their territories birds were challenged persistently by other residents. Defense by the resident male gradually diminished as he ceased to chase the invader consistently, and eventually permitted the latter to sing from song perches he had formerly controlled. Finally, the invader chased the resident from his own song perches, and the loser soon abandoned part or all of his holdings.

Another factor that may affect dominance is a male's capacity to remain on territory continuously, which is influenced by the amount of time he must spend feeding elsewhere. This in turn depends on the distance to and quality of the flowers on his foraging route. Because a male must return from his early-morning

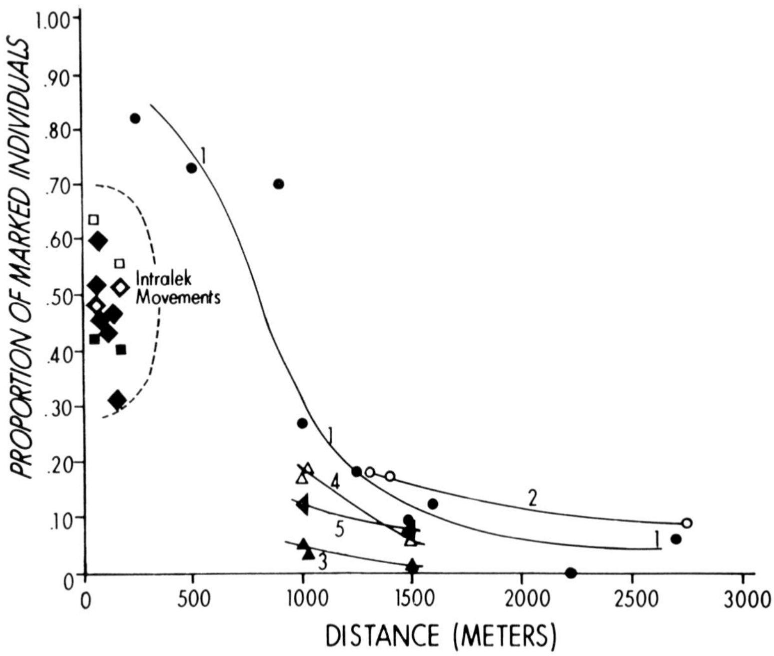

Fig. 25. Proportions of *P. superciliosus* marked at one site that were recorded at other sites different distances away. 1 = movements between leks and foraging areas by adult males; 2 = movements between foraging areas by adult females; 3 = movements between leks by adult males; 4 = movements between leks by immature males; 5 = movements between leks by adult females. For intralek movements, squares = lek I; diamonds = lek II; solid figures = adult males; open figures = immature males. See Figures 3 and 6 for locations of the different sections of leks I and II, respectively.

foraging period or "breakfast break" with enough accumulated food reserves to permit intense activity over the next 2–3 hours, the length of his breakfast break may be an indicator of the quality of his foraging route. However, no significant correlation exists between lek position and the length of the early morning foraging period (Table 15). Being able to spend less time foraging and hence presumably more time on the lek, apparently does not increase the probability of holding and maintaining a central territory in *P. superciliosus*.

Inter- and Intralek Movements

Movements between leks by adult male *P. superciliosus* were extremely rare (Fig. 25). In 4 years, we recorded just 2 of 68 residents of leks I and II appearing on the other lek. OW, a resident on lek II, was captured on lek I at the end of the 1971 lekking season. WY, marked on lek I in February 1971, was a resident of lek II in April 1971. Only one of 33 known residents of lek I, OY, was captured on lek III, and we recorded no movements in the other direction despite intensive netting on and around lek I. In much less extensive netting we recorded one resident of lek II on lek III (GBP), but movement between these two leks may well be slightly higher as both are located on the Quebrada El Salto, probably a major flyway (see Fig. 1). We never saw a known resident of one lek interacting with residents of another lek, and it is difficult to see what advantage he would gain by doing so.

Although there are few available data, there is some suggestion that nonresident adult males, presumably birds without a lek territory, moved between leks much more often. One such male, YR/OG, was recorded on leks Ia, Ib, and II between 1969 and 1970. In 1972, BOP was recaptured on leks I and III. These males may have tried to establish territories on one lek after failing to do so on another, or they may even have tried to establish on two leks simultaneously. In either case the long-distance movements must be expensive in terms of time and energy. A nonresident making a shift of approximately 1 km between leks must either fly a long distance to reach his old foraging route, or must establish a new foraging route with each shift. This may explain, in part, why we so rarely recorded a nonresident male in more than one season.

During 1972 and 1973 we obtained data on movements between leks I and III. Seven young males were marked or seen on one lek at the end of a lekking season and were found to be residents on the other lek the following season. In five of these cases, young males marked on one lek definitely sang or interacted at the end of the season on the other lek where they eventually became residents. Thus, a period may exist at the end of the lekking season during which young males move from lek to lek until they finally establish themselves on one. Such movements would be least expensive at this time of year because the number of major food plants in good bloom was at or near its annual maximum (Fig. 8, Table 4; Stiles 1975, 1978a). This behavior probably was important in giving young males a second chance should they fail to gain a territory on a given lek. As noted earlier, a young male that obtained a territory at the end of a lekking season was virtually assured of resident status on the same lek in the next season. No young male ever gained a territory on one lek, then became resident on another. This period of juvenile male dispersal at the end of the lekking season doubtless promotes gene flow and reduces the inbreeding potential of each lek, since established adult males never change leks (see Buechner and Roth 1974).

Females occupied mainly the areas between leks for nesting and foraging (Table 9, Fig. 1). Thus it is not unlikely that a female ready for copulation may have had to choose from two or more approximately equidistant leks, and she could well visit more than one of them during a breeding season. Females might therefore be seen or captured on two or more leks relatively frequently. Because females so seldom visited leks our data are rather limited. We recorded only 4 of 44 females on more than one lek. However, a more appropriate index may be the ratio of the number of times females were resighted or recaptured on the same versus different leks. This indicates a higher degree of movement as females were recaptured or resighted 12 times on the lek where they were first seen or marked, and 4 times on different leks.

Intralek movements by resident males were consistently more frequent than interlek movements. From 30–60% of the residents of a given sector of a lek were seen or recaptured on another sector of the same lek (Fig. 25). The proportions of birds moving between leks Ia, Ib, and Ic were similar to the proportions of birds moving between sectors of a lek that are similar distances apart; this supports our conclusion that all subdivisions of lek I are part of the same lek unit. However, the fraction of birds seen or captured on more than one sector of the same lek was surprisingly low, given the distances involved, which are much shorter than the average foraging distance (Fig. 11). As expected, young males showed higher proportions of such movements than did adults, about 60% vs. 40% for lek I (Fig. 25). Territorial shifts between lek sectors occurred rarely (e.g., RR, GBY, and BW discussed in the preceding section). Also, some young males that established territories on one sector at the end of a lekking season became residents of another sector in the next year (e.g., YB). A complicating effect that is not discussed in detail here is that of song type. Males on different sectors of a lek often had different songs, and young males, once they had acquired a song, tended to settle in the sector with males having the same song (Wolf and Stiles MS).

The amount of interlek movement by males can best be evaluated in the context of movements of these males from leks to foraging areas or of females between leks and foraging areas (Fig. 25). The proportion of males moving over a given distance from one lek to another was far smaller than that moving the same distance but from a lek to a foraging area. The same held for intralek movements. The proportion of males moving to a foraging area was far higher than that moving to another sector of the same lek the same distance away. Females showed rates of movement similar to those of lek males to foraging areas and of young males between leks (Table 8, Fig. 25). The low rate of interlek movement by adult males is explicable in terms of time and energy. The rivals most likely to displace a territorial male are his neighbors on the same lek. Residents of other leks represent no threat whatever to his territory, because interlek transfers by territorial males virtually never occur. Residents of other sectors of the same lek are at most a minor peril, as intersector transfers are rare also. A resident male gains little by flying to another sector of the lek and interacting with the residents there, and gains nothing by going to another lek. Interlek movements represent a considerable waste of time and energy that could be better employed in reinforcing the male's dominance position with respect to his most frequent challengers and intruders, the neighboring residents of his own lek. This offers a male his best chance to expand or shift his territory, without leaving it unguarded very long in an interaction.

DISCUSSION

Lek Evolution in *Phaethornis superciliosus*

Leks of *P. superciliosus* occur in the densest part of a very dense habitat, the understory of disturbed, tropical wet forest. Residents on their territories are rarely in visual contact, an obvious and important difference from the residents of leks of certain grouse and shorebirds. The form of the displays of *P. superciliosus*, including their orientation about a song perch, obviously has been influenced by the structure of the habitat. It is unlikely that these leks evolved as an antipredator device analogous to a flock in enabling more rapid detection of predators (Lack 1968; Trivers, pers. comm.). This presumed advantage of leks should hold primarily in habitats sufficiently open so that several residents simultaneously can observe the approach of a predator. However, singing from dense thickets may give some protection to individual males.

Our limited evidence indicates that the lek territories of male *P. superciliosus* are the sole sites at which mating sequences are initiated (although the actual mating can occur elsewhere following a chase from the territory). Lek residents therefore should enjoy a strong mating advantage over nonresidents, either non-territorial males or males that display solitarily. Several lines of indirect evidence suggest that male *P. superciliosus* can obtain more matings on a lek than they can by displaying alone. First, males apparently preferred to hold territories at or near centers of activity, as the cases of BW and GR mentioned above clearly indicate. Second, if males could achieve significant mating success by displaying solitarily, one would expect to find lone males scattered throughout the forest. That such males did not occur (in spite of the small number of leks and the fact that some males failed to obtain lek territories) supports the thesis that males can produce more offspring by joining a group. If this is so, then the lek system generates a positive feedback reinforcing the tendency for males to compete for lek territories rather than to establish solitary mating stations.

What then limits lek size? Why do not all males converge on a single large lek? The most likely limiting factor is the distribution of suitable flowers. This could make it energetically unfeasible for more than a certain number of males to join any given lek as they would have to fly too far to forage. Also, concentrated foraging of too many males could create a wide "denectarized zone" around a lek that would be energetically expensive for a female to cross to reach the lek, and females might accordingly choose to visit smaller leks. Another possible limitation on group size is that an incoming male probably stands a better chance of becoming a resident (and hence achieving copulations) of a small lek than of a large one, especially given the short average lifespan of *P. superciliosus* that makes moving up the dominance hierarchy with age unlikely.

Within the lek system females should attempt to mate with the male(s) most likely to provide their offspring with the greatest chance to survive and reproduce, and selection should favor those capable of making such discriminations. A possible indicator of a male's fitness is his proficiency in territorial defense—in effect, his dominance. Our data indicate that peripheral territories tended to be held by younger males who presumably were subordinate and would not predictably contribute as much to a female's fitness as the more dominant males on a lek which held the most centrally located territories. If mating success is correlated with dominance in male *P. superciliosus*, then selection should favor the tendency to hold central territories as these are the most strongly contested territories on the lek and their holders should be the most dominant males present.

The essential feature of a central territory is probably not its location *per se*, but rather that it is surrounded by other territories, with defended boundaries on all sides. Central territories also tended to be in the densest vegetation on the lek. The net result of these factors is that central territories were usually the smallest and closest together. This, coupled with the fact that central males sang more persistently during the period of female visitation, created a gradient of activity or overall song volume from the center of the lek (or section thereof, in a large and dispersed lek like lek II) outwards. This gradient was maintained partly through the aggression of central males toward peripheral ones, resulting in less uninterrupted singing during female visits to the latter.

We have no evidence that female *P. superciliosus* remained on the lek for long periods, although they might make several visits before mating. This, plus the lack of any morphological characteristic correlated with position (= dominance) among males, potentially makes the activity gradient a female's most reliable cue to a dominant male. It may also indicate to a male the direction in which to expand or shift his territory should the opportunity arise. The proximate goal of intermale competition is thus to hold a territory as high in the activity gradient as possible. It is important to note here that activity gradients need not be symmetrical in all directions; due to configuration of lek vegetation or topography, "central" males need not be in the geographic center of the lek. A large lek of *P. superciliosus* may have several "centers" (which often correspond to song dialect groupings), much as do leks of certain grouse and shorebirds (Hogan-Warburg 1966; Wiley 1973, etc.).

Some evidence that females cue on activity centers in mate choice exists for other lek species including the Ruff (*Philomachus pugnax*) and several grouse. Ruffs on large leks tend to keep satellite males off their territories, while males on small leks allow satellite males to co-occupy their territories (Hogan-Warburg 1966; van Rhijn 1973). This could be a mechanism to increase the activity levels on small leks, and males on such territories get more matings. Similar levels of activity could be generated on large leks simply by the larger number of males. In the Greater Prairie Chicken (*Tympanuchus cupido*) females apparently will not submit to mounting by males without the stimulus of the high activity levels of the center of the lek (Robel 1967). For Black Grouse (*Lyrurus tetrix*), Kruijt and Hogan (1967) suggest that density of males and activity levels per unit area may be important components of female choice of a mate.

In this regard, the accretion of young males on the periphery of *P. superciliosus* leks late in the breeding season may be advantageous to the residents. So long as the young males do not establish themselves and become too active and vocal, their chances of mating are slight. However, their very presence adds to the activity levels of the lek, which may enhance its effectiveness in attracting females. As most young males were not permitted to establish stable territories until the adults began to leave the lek, they were unlikely to attract females away from the adults but rather enhance dominance/activity gradients in favor of the latter. The advantage to young males of such precocious lek activity doubtless lay in the experience gained that virtually guaranteed the survivors a position on the lek in the next breeding season (Table 13). Similar arguments have been advanced for the role of young males on leks of some tetraonids (Robel 1967; Kruijt and Hogan 1967).

Although the phenomenon of dominance/activity gradients seems widespread in lek systems, an important structural difference may exist between leks of *P. superciliosus* and those of most grouse and Ruffs. In these latter leks, one or a few dominant males account for virtually all of the matings; subordinate residents have very little chance of mating except insofar as they can move up the dominance hierarchy with time. A nonterritorial male is only slightly worse off than a subordinate resident in terms of his immediate chances of mating, but the latter has a better chance in the future if he survives.

The dominance/fitness gradient on a *P. superciliosus* lek appears far less steep, such that peripheral males stand at least some chance of mating. The major gap in fitness is probably between territorial and nonterritorial males, with less of a difference between dominant and subordinate residents. This may explain why returning adult males almost invariably try to regain their old territories instead of taking up a more central, vacant territory. Subsequent centripetal shifts occur during the lek season, with the male using his territory as a base from which to occupy adjacent areas, but never risking loss of his old area before gaining control of the new. The dominance hierarchy on a *P. superciliosus* lek probably is less rigidly structured spatially than, say, a grouse lek. This is a logical consequence of the high annual turnover of lek residents so that males rarely survive to move up the dominance hierarchy with age.

The hypothesized dominance/fitness structures of *P. superciliosus* and grouse leks are diagrammed in Figure 26. A mating advantage of the center will set up a genetic feedback that will tend to drive a *P. superciliosus*-type lek system to evolve toward a more structured, grouse-type system unless other factors intervene. Factors limiting such a feedback on *P. superciliosus* leks probably include not only the high male turnover, but also a generally lower probability of mating due to the dense lek vegetation, the rarity and unpredictability of female visits, and the fact that all males frequently must leave the lek to feed. That the residents are not in visual contact makes it less likely that a dominant male will be able to interfere with a mating on a subordinate's territory. Dominance on the lek is apparently unrelated to how long the males are away foraging; hence, dominant males have as great a chance of missing a female visit for this reason as do subordinate males.

Unlike certain other hummingbirds that join singing assemblies (Skutch 1958, 1967), male *P. superciliosus* did not defend food sources on or near their lek territories. Indeed, we think it best to specifically exclude from the definition of a lek those territorial systems in which the territories serve other functions as well as that of mating. Spatial separation of defended mating sites and regular feeding areas seem to characterize other bird species that form leks (D. Snow 1962, 1963; Lill 1976). Defense of feeding areas is precluded by the time and energy demands of maintaining a lek territory. Food sources or foraging zones of lek species are usually fixed in space, enabling easy location and rapid visitation, but are generally not energetically profitable to defend, either because of uniform abundance over wide areas (e.g., buds and leaves), rapid turnover and shifting abundance (fruit), or occurrence in small, isolated patches, as in the nectar supplies used by *P. superciliosus*.

The dense vegetation on *P. superciliosus* leks doubtless places a strong selective premium on effective auditory communication. The loud, persistent song, often audible for 100 m in the forest understory, was the sole means of advertisement over any distance by lek males. Visual displays assumed primacy only at very close range, within about 1–2 m. A number of other lek systems in dense vegetation rely on sound, either vocal or mechanical, for distance communication. Examples include the Hammerhead Bat (Bradbury, pers. comm.), manakins (e.g., D. Snow 1962; Lill 1976), cotingas (Gilliard 1962; B. Snow 1972), and other hummingbirds (e.g., Davis 1958; Skutch 1964a; B. Snow 1974).

Visual displays given by *P. superciliosus* appeared to differ little in male–male

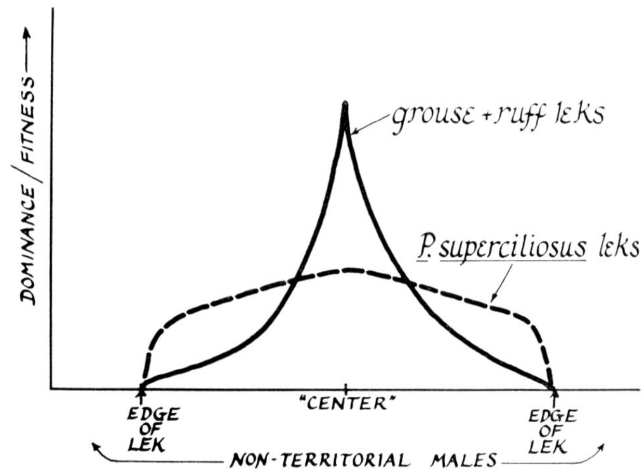

Fig. 26. Hypothesized relationship of dominance and/or fitness to lek position in leks of grouse and Ruff vs. leks of *P. superciliosus*. In the former, subordinate, peripheral males have virtually no chance of mating; the *P. superciliosus* system is much looser, with any resident male having a fair chance of mating. "Center" refers to areas of high lek activity, not necessarily to the geographic center of the lek.

and male–female encounters, perhaps reflecting the lack of sexual dimorphism in this species. Without morphological or auditory cues to the sex of an intruder, a resident male should respond aggressively to any conspecific entering his territory and establish his dominance. An intruding male would then leave, while a female could signal her readiness to mate by remaining perched and allowing the male to mount. Because the cost to a male of sperm lost from an inappropriate copulation is minimal, the brief and unpredictable appearances of females probably place a strong selective pressure on the male to attempt mating with anything that sits still long enough for him to mount.

Hermits in general show little or no sexual dimorphism; even in the most dimorphic species, young males pass through a female-like plumage (B. Snow 1974). This situation is very unusual among lek species, in most of which strong intermale competition has produced distinctive, even bizarre male plumages. A frequent correlate of sexual dimorphism in polygynous or promiscuous species is sexual bimaturism (Wiley 1974), that is, different rates of sexual maturation in males and females (usually slower in the former). The high annual mortality rate of male *P. superciliosus* probably reduces drastically the potential mating advantages of delayed maturation and instead selects for lek residency at the earliest possible age.

Sexual dichromatism in hummingbirds often is related to purely aggressive interactions, frequently involving territoriality at flowers. In highly dimorphic species, females usually are nonterritorial and males territorial (Wolf 1969, 1975b; Stiles 1973). In several monomorphic, brightly colored species, both sexes hold territories during at least part of the year (Wolf 1969, 1975b; Stiles and Wolf 1970). Monomorphic but dull-colored hermits lack territoriality at flowers. Iridescent colors apparently are not important aggressive signals in hermits, perhaps because of low light intensities on display areas. Many nonhermit lek hummingbirds are highly dichromatic, suggesting that color is useful for aggressive signalling in at least some of their displays.

The daily temporal pattern of lek activity in *P. superciliosus* is organized around at least three factors. First is the infrequent and unpredictable appearance of females, leading to strong selective pressures for males to spend as much time as possible on their territories, especially during the late morning when females are most likely to appear. Second is the recurring necessity for defense of the lek territory and reassertion of dominance position on the lek. Most of this was compressed into a short period of aggressive activity just after dawn, although nonresident males could intrude at any time. A resident should leave his territory unguarded as little as possible because an intruder that has time to sing and interact on the territory may be much more difficult to expel than one who has not.

The third major factor is the availability of nectar, which follows a highly predictable daily course around which all other activities must be organized if the most favorable balance between foraging and lek activity is to be attained. The timing of the breakfast break corresponded closely to that of maximum daily production and accumulation of nectar in important food flowers (Fig. 9; Stiles 1975 and unpubl. data). The burst of aggressive activity before this break probably reflected the importance of reestablishing dominance position on the lek as early as possible, and redefining territorial rights prior to the arrival of females. Theoretically, a nonterritorial male could try to establish himself in a lek territory during the breakfast break when the resident was absent, but he would be sacrificing the best foraging time of the day to do so and would be at an energetic disadvantage in the subsequent confrontation with the returning resident. That females appeared on the leks mostly in mid- to late morning probably also relates to nectar availability. It is at this time that they should have the most energy reserves and be best able to leave their foraging areas to engage in courtship. The less synchronized activities of residents from midmorning on probably reflected varying distances flown and rates of energy intake on feeding bouts. Food supply does not appear to play a major role in determining the daily schedule of leks in other bird groups, doubtless because no distinct daily rhythm of availability exists for fruits, buds, etc. On the other hand, predation may strongly influence daily schedules in at least some grouse leks (Hartzler 1974) but seems to be of little or no importance in *P. superciliosus* leks.

The seasonal pattern of lek activity in *P. superciliosus* seems closely tied to the blooming seasons of certain food plants, especially several species of *Heliconia*. Food supply can regulate the lekking season either directly through the males' energy budgets, or indirectly through the nesting—hence mating—season of females. In *P. superciliosus* the onset of lekking appears related to the first factor, the end of the lek season to the second. A similarly close relationship between breeding activity and blooming of food plants exists in several other hummingbird species (review in Stiles 1973), including *P. guy*, a lek form (B. Snow 1974). The lekking seasons of manakins in Trinidad appear to correspond to periods of fruit abundance, though the relationship may not be as precise as that between flowers and hummingbird leks (D. Snow 1962, 1963). Comparatively little quantitative information is available on the effects of food supply on lekking seasons in other species; for most of these, detailed observations on foraging patterns have yet to be made.

Comparisons with Other Hermit Species

The subfamily Phaethorninae, collectively called hermits, comprises 6 genera and 31 species (Meyer de Schauensee 1966). In the large genus *Phaethornis*, the species *superciliosus*, *longuemareus*, and *guy* form leks throughout their ranges; *ruber* forms leks in some parts of its range and not in others (Davis 1934; Oniki 1970), while some species (e.g., *pretrei*) apparently do not form leks at all (Ruschi 1950). Leks sometimes occur in the genus *Threnetes*, but this varies even in the same species at the same locality (Skutch 1964a; B. Snow 1973b). At La Selva, we found solitary male *T. ruckeri* singing in some areas, and leks of three to six birds in others. In *Glaucis*, leks are not known. In Trinidad, *G. hirsuta* males hold individual territories, within which females nest (B. Snow 1973a; Snow and Snow 1973). At La Selva we found no suggestion of lek behavior in *G. aenea*. In *Eutoxeres* we observed a small lek (3–4 birds) at La Selva. The social systems of the remaining species and genera of hermits apparently are unknown. Thus, lek social systems are common, but by no means universal, among hermits but seem particularly characteristic of the genus *Phaethornis*.

Lek residents in all lekking hermits for which sex has been determined invariably are males (Arp 1957; D. Snow 1968; present study). Females visit the leks for the purpose of mating. The final phase of mating, copulation, may occur inside the male's lek territory or elsewhere, sometimes around the nest of the female (B. Snow 1974), following a chase starting from the territory. There has been no suggestion of matings initiated away from the lek in any of these species. In all species most lek activity is between males, and visits by females are relatively rare.

Different hermit species vary considerably in the temporal continuity of leks. Leks of *P. superciliosus* have been found at the same site for at least 12 years in Guyana (Davis 1958) and 8 years in the present study. Leks of *P. longuemareus* and *P. guy* also may occupy the same site consistently for several years (B. Snow 1974; D. Snow 1968; Wiley 1971; pers. obs.). On the other hand leks of *Threnetes* and *Eutoxeres* are much more variable. A lek of *Eutoxeres* at La Selva was active between December 1971 and February 1972 but not the following year. A 3–5 bird lek of *Threnetes* remained more or less constant in size and position for 3 years; in another site we found a solitary male in some years and a lek of up to six males in other years. Several calling sites have been used consistently by solitary males for up to five years; turnover of individuals was not studied.

Leks of several species are located near flyways and/or feeding areas. In Trinidad, leks of *P. guy* are found on steep slopes or ridges where streams are available as flyways, and where a major food plant, *Heliconia bihai*, is plentiful. Sometimes leks even are located in stands of *Heliconia*, which is exceptional in *P. superciliosus*. However, no large, dominant, nonhermit species like *Chalybura*, that might disrupt leks by defending *Heliconia*, occurs in Trinidad. In neither Trinidad nor Costa Rica are *P. longuemareus* leks especially near flowers. This species visits many small, largely insect-pollinated flowers and also eats relatively large quantities of insects (B. Snow and D. Snow 1972, pers. obs.); specific foraging sites may be less important than in the other lek hermits. *Glaucis* in Trinidad is a nonlek species territorial along streams in areas important for nesting and foraging (B. Snow 1973a). No hermit is known to consistently defend flowers.

In males of all hermits studied so far, the main advertising technique whether alone or in a lek is to sing. In some larger species of *Phaethornis*, like *superciliosus* and *guy*, the song is simple and repeated monotonously. At the opposite extreme is the complex, rather melodious song of *P. longuemareus* (Skutch 1951; D. Snow 1968; Wiley 1971). *P. ruber* has a song intermediate in length and complexity (Davis 1958; Oniki 1970; B. Snow 1973b). The large *P. yaruqui* also has a notably loud and complex song (Ruschi 1961). *Threnetes ruckeri* has a long relatively melodic song (Skutch 1964a; pers. obs.), while *T. leucurus* has a simpler song, a three-note phrase rapidly repeated, in the manner of the large species of *Phaethornis* (Snow 1973b). *Eutoxeres* has a simple song consisting of a single, repeated note, but the successive notes differ in tone and pitch (pers. obs.). Although possessing a song comparable to that of *P. ruber* in complexity, *Glaucis hirsuta* rarely sings. Perhaps the unusual social system and restricted habitat of the species in Trinidad make vocal communication by males largely unnecessary (B. Snow 1973a).

Local song types exist in *P. longuemareus* (D. Snow 1968; Wiley 1971), *P. superciliosus* (this study) and probably in *P. guy* (B. Snow 1974). Song varies geographically in *P. superciliosus* (B. Snow 1973b), *P. longuemareus* (comparing songs in Costa Rica with those reported by D. Snow 1968), and *P. ruber* (Davis 1958; Oniki 1970). In species with local song types considerable geographic variation is expected because the final form of the song is learned by young males (B. Snow 1974; Wolf and Stiles MS).

Members of *Phaethornis*, *Threnetes*, *Glaucis*, and *Eutoxeres* all have conspicuous tail patterns, and characteristically wag the tail up and down when perched; singing males wag the tail very strongly. Some authors (e.g., Slud 1964; Skutch 1964a) suggested that the tail wag is coordinated with the rate of calling, but this was not so in *P. superciliosus*. The rate and intensity of the wag increase in all species, however, and the tail often is fanned when a conspecific approaches or enters the singing male's territory.

All hermits so far studied have some type of hover display in which the displaying bird hovers before a perched bird (sometimes both may hover), displaying its gape and/or throat pattern, depending on the head and bill coloration. *P. guy* has a conspicuous gape, but the adult has no conspicuous face pattern; it has a specialized gape display. *P. superciliosus* has a fairly conspicuous gape and a striking face pattern and displays both of these in the float. *P. guy* has no display strictly analogous to the float, but, as discussed below, the *tock* display contains elements of both gape and bill-pop and float. Gape displays are not known in *Glaucis* or *Threnetes*, the species of which have dull mouth linings; instead the bright cinnamon breast (*Glaucis*) or conspicuous throat pattern (*Threnetes*) is displayed by the hovering bird (B. Snow 1973a, b). The larger *Phaethornis* species and *Threnetes* hover with the long axis of the body vertical, but in *Glaucis* and the smaller *Phaethornis* the body axis is horizontal, and the tail is raised—the so-called "boat" posture (Skutch 1951). A back-and-forth movement is most typical, but *P. longuemareus* and *P. ruber* may turn complete circles in some situations (Skutch 1951; Arp 1957; Davis 1958). In general, the aerial displays of these small species are considerably more elaborate than those of the larger hermits, including additional sound effects (Davis 1958; B. Snow 1973a). Interestingly, perched displays such as side-by-side have not been reported in the small hermits.

The extension of the tongue plays a role in the hover displays of *P. ruber* (Davis 1958) and *Glaucis* (B. Snow 1973a). In a Guyana population of *P. superciliosus*, B. Snow (1973b) observed that the bird visiting a territorial male typically perched beside him and touched the inside of his displayed gape with its tongue. We never observed this in the Costa Rican population of *P. superciliosus*, although the display otherwise seems similar to a side-by-side. Guyanan populations of *P. superciliosus* also have a special flight call used only on the lek; we never heard such a call in Costa Rica. It is worth noting that the conspecificity of the Central and South American birds is in question (Meyer de Schauensee 1966).

Tock or *pop* sounds are produced by *P. guy*, *P. superciliou*s, and rarely, by *Glaucis hirsuta* (B. Snow 1973a). The associated gape display is more elaborate and spectacular in *P. guy* than in *P. superciliosus*, involving a more conspicuous spreading of the mandibular rami; the gape is bright red in *P. guy*. The tock display of *P. guy* has both side-to-side and darting movements; the *tock* or *pop* is given at the ends of rapid sideways arcs. In addition to using the *tock* in all of the same situations as *P. superciliosus* use the bill-pop, male *guy* also characteristically *tock*-display when changing perches in their territories, or may hover and *tock*-display repeatedly over a perch. *Tock* or bill-pop sounds may be produced by a sudden expulsion of air from the glottis in conjunction with a snapping open of the bill. B. Snow (1974) described several cases of loss of voice in *P. guy* and suggested that the *tock* is produced vocally in that species because on losing its voice, a bird no longer could *tock*. Because air coming from the lungs or air sacs must pass through the syrinx en route to the glottis, the two explanations are not mutually exclusive.

Perch exchange sequences are known for *P. superciliosus*, *P. guy* (B. Snow 1974), *P. longuemareus* (D. Snow 1968), and *Threnetes leucurus* (B. Snow 1973a); they probably occur in most hermits as they seem to be a ritualized form of precopulatory mounting that occurs regardless of the sex of the sitting bird. Copulation in most or all species resembles that in *P. superciliosus* (Fig. 7H); Davis' (1958) report of aerial copulation in *P. ruber* has not been confirmed by other authors (B. Snow 1973a).

Several authors attempt to discriminate between displays given to females and those given to other males by lek residents. In the sexually dichromatic species (*P. ruber*, *P. guy*) this may be feasible, but it is complicated by the fact that in at least the latter, young males are indistinguishable from females by plumage and approach the lek like females (B. Snow 1974). Many of the supposed female visits to the lek ending in chases may have been by yearling males that had not yet obtained a lek territory. In view of the variability in display sequences found in *P. superciliosus* (Table 3), there is no satisfactory evidence that the displays of a lek resident to a visiting male or female are qualitatively different, and evidence for even a quantitative difference is weak (B. Snow 1974). The similarity may include even mounting and attempted copulation if a perching male sits still. Homosexual mounting and similarity of mating and male aggressive displays in *P. superciliosus* are consistent with the limited information for other hummingbirds, including nonlek species (Stiles and Wolf 1970; Stiles 1973; Wolf 1975a; Stiles and Ortiz-Crespo MS).

The brief, rare, and unpredictable appearances of females on male territories (B. Snow 1974; this study) are probably major factors in the frequency of leaf copulations or false matings in most hermits and many other hummingbirds. Such behavior has been reported for various species of *Phaethornis* (Arp 1957; D. Snow 1968; B. Snow 1973a, 1974; present study), *Threnetes* (B. Snow 1973a), *Glaucis* (B. Snow 1973b), and has been seen in *Eutoxeres* (Stiles, pers. obs.). The analogy to masturbation in mammals is obvious.

The only other study of lek behavior in hummingbirds that deals with such parameters as survival, recruitment, molt, and foraging of lek residents is the work of B. Snow (1974) on *Phaethornis guy* in Trinidad. *P. guy* is a fairly close relative of *P. superciliosus* but shows marked sexual dimorphism, and young males pass through several more or less intergrading plumage stages before acquiring the definitive plumage. B. Snow did not work with marked birds, but stated that "differences in head markings, combined with differences in calls (songs), made it possible to identify a number of individuals with a high degree of certainty." Her interpretation of plumage sequences, which is crucial to her evaluations of survivorship and recruitment, is based largely on territorial occupancy by birds of similar song type in different years. In *P. superciliosus*, at least, it may not be safe to assume that a bird singing from the same territory with the same song as the previous years' resident is in fact the same individual. We have many cases of individual turnover in which the song type of the territory holder remains the same (see Fig. 6, esp. BRG and WPB). Therefore, certain of Snow's conclusions require confirmation from studies with marked birds.

In general, temporal and spatial patterns of lek activity in *P. guy* resemble those of *P. superciliosus*. Territory size in *P. guy* averaged slightly larger, but from B. Snow's comments regarding frequent visual contact of singing males we infer that the vegetation on the *guy* leks was somewhat more open than that on our *superciliosus* leks. The lekking season of *guy* began in November, but less suddenly than that of *superciliosus*; only by mid-December were all males attending the lek regularly.

Highest display intensities in *guy* (number of *tocks* per minute) were attained at the start of the lekking season and declined slightly thereafter, as in *superciliosus*. The daily patterns of lek activity of individual male *guy* fell into late morning and afternoon periods as in *superciliosus*, but no data are available for *guy* before 0630, so an early morning activity period is conjectural. Most adult male *guy* regularly attended the lek from late December through May or June. *P. guy* and *superciliosus* differed greatly in the timing of the annual molt. In *guy* the molting season followed the lekking season with little or no overlap; adult males left the lek for the year as they started to molt. The slow molt and possibly unique timing mechanism of *superciliosus* (Stiles and Wolf 1974) result in extensive molt-breeding overlap and correlate with a somewhat longer lekking season than in *guy* (8–9 versus 6 months).

Most young *guy* males first appeared on the lek in April or May and moved into central territories as the adults left; their behavior closely resembled that of young male *superciliosus*. At this time they were in fresh plumage, and Snow concluded that they had just completed a molt and must, therefore, be in their second season. From our experience with *superciliosus*, including aging of young males by bill corrugations, we think it more probable that the fresh plumage of young male *guy* was the first-year plumage, and that these males were only a few months old. B. Snow stated that these young males went through a second season in this same plumage. If our interpretation of their ages is correct, these young male *guy* wore their juvenal plumage for 15–18 months and molted annually thereafter. Precisely such a molt schedule has been documented in *P. idaliae* (Ruschi 1967). In *P. superciliosus* the juvenal plumage also is worn for slightly longer than any succeeding feather coat (Stiles and Wolf 1974).

B. Snow (1974) distinguished three other plumages of young males, which she interpreted as different year classes. These classes appear to us to intergrade and may even represent only a single, variable second-year plumage. Thus the adult male plumage in *P. guy* is attained somewhere between the second and fourth complete annual molt; this is the most elaborate plumage sequence yet known for the genus *Phaethornis*.

According to B. Snow, annual turnover of lek males in *P. guy* was 10% or less, a far lower figure than we found for *P. superciliosus*. Part of this difference could reflect inherent biases in our respective methods. The use of territory location and song type to identify individuals certainly is unsafe in *P. superciliosus* and might well underestimate turnover in *P. guy*. On the other hand, our marking system could have reduced survivorship in *P. superciliosus*. However, most mortality occurred when we handled the birds least, and we excluded from the analysis all individuals that we felt might have been adversely affected by the tags. Conceivably, the tag could weaken a bird just enough to affect its survivorship over the lean season, but such effects were probably very minor at best. Each tag weighed less than 0.1 g, far less than a crop of *P. superciliosus* when filled with nectar (Hainsworth and Wolf 1972b), and the plastic was too stiff and short to become tangled easily in vegetation. The tags may interfere with nesting by females (Waser and Calder 1975), but our data on survivorship treat only lek males.

Even allowing for these possible errors of estimation, we feel that a real demographic difference exists between *P. guy* on Trinidad and *P. superciliosus* at La Selva, reflecting differing rates of both reproduction and mortality. The longer breeding season of *P. superciliosus* makes possible more nesting attempts. Indeed, there is time for up to three successful nestings in the breeding season of *P. superciliosus*, two in the breeding season of *P. guy*. Thus the annual reproductive output of the former could be as much as 50% greater than the latter, although the difference probably is less due to the low nesting success in both populations. Regarding mortality, at least one important predator on adult hummingbirds (Stiles 1978b), the Tiny Hawk (*Accipiter superciliosus*), is absent from Trinidad (ffrench 1973). However, most other potential predators, including snakes, are well represented on Trinidad (D. Snow and B. Snow 1964). More important than differences in predation between the two areas may be differences in the seasonal distribution of nectar resources. In Costa Rica the season of low nectar availability from late October to early December was very severe, with no abundant food plant available over most of La Selva. The relatively short hiatus between the blooming season of the two major food plants of *P. guy* on Trinidad was bridged by the blooming of a third species (B. Snow 1974). To judge from the number of feeding records for the latter, it was fairly abundant (B. Snow and D. Snow 1972). Without the high mortality during October and November, annual survivorship in *superciliosus* at La Selva would approach that claimed by B. Snow for *P. guy* on Trinidad.

If one accepts B. Snow's estimates of turnover in *P. guy* and considers the difference in reproductive output of the two populations to be 50% or less, then there must be a relatively greater surplus of young males produced each year in *P. guy* that do not obtain lek territories. A possible indication of this is the high frequency of "female" visits to *P. guy* leks. Snow did not consider the fate of such males, or of the demographic consequences of such a low recruitment rate. The presence of numerous nonbreeding males may limit reproduction of the population by depleting nectar supplies that otherwise could be available to females. It would be most useful to know the interval between successive clutches in females of the two populations, but such data are virtually impossible to obtain in such a dense habitat.

Evolution of Lek Behavior in Hummingbirds

Two conditions seem necessary for the evolution of lek behavior in any organism: (1) the emancipation of the male from participation in the reproductive effort beyond insemination and (2) sufficient time beyond that required for maintenance to control a lek territory (D. Snow 1962, 1963). A variety of animal species, including vertebrates and invertebrates, meet the first condition. For birds and mammals, Orians (1969) suggested ecologic and demographic situations in which to expect male emancipation. Among hummingbirds, almost all species meet the prerequisites for the evolution of lek behavior, apparently related to the fact that the clutch is two eggs throughout the family (Van Tyne and Berger 1959; Orians 1969). However, several instances of apparently secondary reestablishment of male aid in reproduction have been discovered recently (Wolf and Stiles 1970; B. Snow 1973a). Male aid, however indirect, is probably incompatible with the time and energy demands of a lek social system. One therefore should not expect leks to occur in species in which the females can produce more offspring with aid from the male than without it (cf. Orians 1969).

Once emancipated from the reproductive effort, male hummingbirds are under selection chiefly to increase mating frequency and reduce maintenance costs. Males of many species combine feeding and mating by holding flower-centered territories during the breeding season. Generally, male hummingbirds are dominant to females and control the best feeding areas, so females attempt to forage on male territories. Most matings are initiated on these territories as a result of foraging by females (Wolf and Stiles 1970; Stiles 1973; Wolf 1975a). The number of flowers a male can control on such a territory affects not only his energy budget but probably his mating success as well; the flowers in effect constitute a secondary sexual character (Wolf and Stiles 1970; Stiles 1973).

These mating and feeding territories and lek territories, representing complete coincidence and complete separation of defended mating sites and feeding areas, obviously are but the ends of a continuum. Within a single species some males control enough flowers on territory to satisfy all of their energy requirements, and others that hold fewer flowers are required to forage outside of their territories for varying periods (Stiles 1971a, 1973). In fact, in *Calypte anna* of California, changes in the dispersion of flowers can produce local social systems that run the entire gamut from food-centered mating territories to essentially leks. In the latter case, males hold flowerless territories in chaparral vegetation and commute to *Eucalyptus* trees to feed (Stiles 1973).

In view of time and energy, the optimum strategy for a male would be to hold a food-centered mating territory rich enough to support him for a day. This would enable him to forage efficiently and to be present continuously so as not to miss visits by females. Whether a male can hold such a territory depends on the abundance and dispersion of suitable flowers, the number of competing hummingbirds, the male's dominance status among his conspecifics, and the position of his species in the interspecific dominance hierarchy (Stiles and Wolf 1970). When it is consistently impossible for males of a given species to hold such rich breeding-feeding territories, selection may favor the evolution of lek mating systems. Certain advantages of flower-centered territories with regard to the criteria available to females for male choice are also offered by lek systems, and these may not be offered by other alternative strategies, such as males displaying solitarily. These include the fixed, constant locations of male mating stations and an index of male's dominance status independent of his personal characteristics —his position in an activity gradient. Lek social systems and food-centered territories should thus represent adaptive peaks, and one may expect that more hummingbird species of a given area show these types of mating systems than other, intermediate types. At La Selva at least, this prediction is upheld. At least 2 species (*Chalybura urochrysia* and probably *Amazilia tzacatl*) show food-centered mating territories, while 6 (*Eutoxeres aquila*, *Phaethornis superciliosus* and *longuemareus*, *Threnetes ruckeri*, *Klais guimeti*, and *Amazilia amabilis*) form leks. *Glaucis aenea* may hold individual territories (cf. B. Snow 1973a). The mating systems of the other common resident species, *Florisuga mellivora* and *Thalurania furcata*, are unknown. The critical determinant of the adaptive peak toward which a species may evolve appears to be whether breeding males can control rich feeding areas which then serve as mating stations.

To understand the evolution of lek behavior it is thus important to identify factors that may prevent males of a given species from holding food-centered mating territories. Lek-forming Costa Rican hummingbirds include (1) subordinate species in the community that are forced from defensible food supplies by dominant species and (2) species that have specialized on a food supply whose spatial and/or temporal distributions make it nondefensible (Brown 1964; Gill and Wolf 1975; Wolf et al. 1975). Included in the latter category are very large, dominant species that use resources defended by subordinate species but for which sufficiently large concentrations of flowers to justify territoriality seldom or never occur. Species in this category include *Phaeochroa cuvierii* (Skutch 1964b; Wolf 1970, pers. obs.), *Campylopterus hemileucurus* (pers. obs.) and, in Guyana, *Topaza pella* (Davis 1958).

The lek-forming hermits, as a group, fit into the category of subordinate species (Stiles 1975, present study). The two nonhermits at La Selva that form leks, *Klais guimeti* and *Amazilia amabilis*, are small, subordinate species that are excluded from most rich nectar sources (pers. obs.). Such species are forced to exploit a wide range of food plants, many of which grow in nondefensible spatial arrangements or have insect-pollinated flowers that produce very little nectar (Wolf 1970; B. Snow and D. Snow 1972; Hainsworth and Wolf 1972a; Wolf et al. 1976). It is important to note that the two categories above are not exclusive and that subordinate status with exclusion from rich nectar sources and specialization for feeding at nondefensible food plants can evolve concurrently. This clearly has occurred to a high degree in the hermits, including *P. superciliosus*, and has led to the coevolution of hermits as route-foraging, nonterritorial pollinators, and a group of flowers ecologically and morphologically specialized for hermit pollination (B. Snow and D. Snow 1972; Wolf et al. 1972; Linhart 1973; Stiles 1975). Clearly the exploitation systems of hummingbirds are intimately related to their social systems; the two are tightly integrated in the ecology of any hummingbird species. We contend that the evolution of leks in *P. superciliosus* and other hummingbirds cannot be adequately understood without detailed consideration of selective forces arising from the foraging strategy, interspecies relations, molt, and resource distribution in time and space.

ACKNOWLEDGMENTS

This study was supported by a Chapman-Naumberg Postdoctoral Fellowship from the American Museum of Natural History to Stiles, and National Science Foundation grants GB-7611, 19200, 40108, and 28956X to Wolf. We are grateful to the Costa Rican office of the Organization for Tropical Studies (O.T.S.) especially J. Campabadal, for logistical support at La Selva. Help in the field was provided by P. Campanella, F. R. Hainsworth, L. Hurd, L. F. Kiff, D. Lyon, S. M. Smith, and the participants of several O.T.S. courses. Weather data for La Selva were provided by R. Chavarría and G. Hartshorn. Helpful discussions and comments on the manuscript were supplied by J. Bradbury, P. J. Campanella, D. H. Janzen, P. Marler, G. H. Orians, S. M. Smith, D. and B. Snow, R. H. Wiley, and D. Windsor. To all of these people and institutions we express our sincere thanks.

SUMMARY

The ecology and lek mating behavior of the Long-tailed Hermit, *Phaethornis superciliosus*, were studied over a four-year period at Finca La Selva, in the wet Caribbean lowlands of Costa Rica. The sexes of this hummingbird are alike in plumage but differ in measurements; all lek residents are males. Most of our observations were made on color-marked birds.

Density of *P. superciliosus* leks at La Selva was about one per km^2. Four leks had 5–6 to 20–25 males during the study period. In undisturbed forest, leks were located in dense thickets along streams, which provided convenient flyways as well as habitat for important food plants. Lek territories were mating stations only, never containing enough flowers to affect the male's energy budget. A territory consisted of a male's song perches and the area between them. The foci of lek activity were the song perches. Territory size varied directly with the openness of the vegetation.

Song was the major means of territorial advertisement; males on territory were rarely in visual contact and spent most of their time singing. Distinct song types or dialects occurred in *P. superciliosus*; males singing a given dialect generally occupied a geographical subunit of the lek, and the number of subunits varied directly with lek size. In spite of turnover of individual males, these subunits tended to retain their integrity from year to year; males on particular territories sang the same dialects as their predecessors.

Visual displays were given only when two birds were in close contact, usually at a song perch. The conspicuous facial stripes, orange mouth-lining and lower mandible, and perhaps the long, white-tipped tail were the most important morphological structures used in visual displays, some of which were associated with distinctive sounds. Allowing for differences in coloration of head and bill, the displays of *P. superciliosus* are fairly similar to those of other hermits. No regular sequence of displays was evident in all interactions, but certain displays tended to precede or follow others with some consistency.

The displays and aggressiveness of an intruding male peaked when he first invaded another's territory. During the ensuing exchange of displays, the resident became progressively more active, in most cases finally driving the intruder from the territory. We found no differences in displays given to male and female intruders, and observed homosexual as well as heterosexual copulations. A female apparently signaled her sex by simply sitting still and allowing the male to mount; a male treated anything that sat still as a willing female. Mating sequences were always initiated in the males' territories, although copulation might take place elsewhere.

Lek males foraged mostly within 200–500 m of the lek, depending on the distribution of suitable flowers. The remaining areas were used mainly by nonresidents, mostly females; all nests were found well away from established leks. *P. superciliosus* foraged mainly at flowers with long, curved corollas that were used infrequently by nonhermit hummingbirds that were dominant to hermits and usually displaced them at flowers. Individual *P. superciliosus* visited a series of small, nondefensible clumps of flowers, apparently along a regular foraging route. The most important food plants of *P. superciliosus* at La Selva were several species of *Heliconia*, especially *H. pogonantha*, and *Costus*.

The lekking season of *P. superciliosus* commenced abruptly in late November or December, most males returning to the lek within a few weeks. Onset of lekking was closely correlated with the onset of good bloom by *H. pogonantha*. Lek activity peaked by January, and high levels were maintained through June or July. The nesting and lekking seasons corresponded closely, but nesting did not peak until February or March, as other dry-season flowers began to bloom profusely. A sharp drop in lekking and nesting occurred in August while flowers were still abundant, perhaps reflecting selection against late nestings. Flower availability declined precipitously after late August, and the annual minimum of lekking and nesting corresponded with a period of extreme scarcity of flowers in October and November.

During most of the breeding season, daily peaks of lek activity occurred in early morning, late morning, and midafternoon. Males arrived on the lek at dawn before feeding, and sang and interacted vigorously for 30–45 min. An hour of concentrated, synchronous foraging followed. The late morning included intense singing and less overt aggressive activity, and in the afternoon males did little more than sing. This decline in active territorial aggression through the day paralleled the decline in nectar secretion of major food plants, and the early morning foraging period corresponded to the time of maximum nectar availability.

Most of the rare, brief visits to the lek by females occurred during the late morning activity period, when all observed copulations also occurred. We estimate that only about one female per day visited a lek of 15 males for the purpose of mating, probably a result of the long, asynchronous breeding season. Males therefore had to be ready for months to mate at an instant's notice, yet the number of matings per male was relatively low. This may explain the males' willingness to mate with anything compliant and their frequent copulations with dead leaves.

Young males hatched early in the breeding season began to appear on the leks in March, but did not become numerous until May or June. At first these young males sang hesitantly, with no fixed dialect and from any position on the lek, and they were frequently displaced by the residents. A few young males established peripheral territories by May or June, but most did not become territorial until July when adults began leaving the lek for the year. Residents were dominant until they departed. By August most territorial, singing males on the lek were young of the year. No single factor could account for the departure of the residents; molt played a role in some birds but not in others. The decline of flowers on a bird's foraging route may also have been important; birds that had to find new feeding areas may have had less time to spend on the lek.

The annual turnover of lek residents in the La Selva population of *P. superciliosus* was about 50%, far higher than reports for other neotropical birds, including *P. guy* in Trinidad. About 75% of the annual mortality occurred during October and November when flowers were scarce. Survivorship of adult and first-year males was similar, perhaps reflecting the pronounced year-to-year variations in flower availability that decrease the value of previous foraging experience to adults. Nearly all recruitment of new residents was from the ranks of yearling males. The more experience in territorial behavior a young male had during his first season, the better his chances of becoming a resident the following year. Young males hatched late in the season and with little previous lek experience also had a chance to become lek residents due to the high mortality of older yearlings or adults.

The most dominant residents on a *P. superciliosus* lek occupied territories near the center, with subordinate individuals towards the periphery. Central territories were the most stable in time and space and the most strongly contested. Residents of the previous year returning to the lek nearly always reoccupied their old territories or vacant territories more toward the center of the lek. Young males returning at the start of their first full lekking season were able much less often to reoccupy the sites on which they were territorial or interacted as yearlings but usually had to settle in more peripheral sites. Most movement towards the lek center occurred during the lekking season as central residents died. However, the dominance status of some individuals changed little with age, even over several years. We found no morphological character that was consistently correlated with dominance, although small, short-billed males often failed to obtain territories.

Because they were often in denser vegetation with defended boundaries on all sides, central territories tended to be smaller and closer together than peripheral territories. Central males sang more than peripheral males during the late morning when females were likely to appear, in part because peripheral males were persecuted more by central males. These factors produced a gradient in activity and overall song volume per unit area from the lek center towards the periphery. Male *P. superciliosus* tried to hold territories where activity levels were high. If, as seems likely, females prefer to mate with the most dominant males, then their most reliable cue to a male's status may be his position in the lek activity gradient.

In general, the dominance/fitness gradient on *P. superciliosus* leks appears to be much less steep than that on leks of certain tetraonids and shorebirds. The proportion of all matings accounted for by central males is probably less, and subordinate males have some chance of mating. Unless other factors intervene, any mating advantage at the center should set up genetic feedback driving a hermit-type lek toward more structured, grouse-type conditions. In *P. superciliosus* these factors include the high annual turnover of residents and the general rarity and unpredictability of mating.

Nearly all hummingbirds satisfy the basic prerequisites for the evolution of leks: emancipation of the male from the nesting effort and a highly predictable food source (flowers). The most energetically efficient mating strategy probably is to control a rich flower supply on the mating territory. Leks tend to evolve in hummingbirds when this is impossible, either because defensible flowers are controlled by more dominant species, or because sufficiently rich, defensible nectar sources do not occur in the species' habitat. As a group, the hermits, including *P. superciliosus*, fall into the first category. They have concurrently evolved a high degree of specialization for exploiting nondefensible flowers, and many such flowers have coevolved morphological and ecological specializations promoting hermit pollination. Leks and flower-centered mating territories represent the ends of a continuum of mating types in hummingbirds. At La Selva most species show one extreme mating system or the other; relatively few species are intermediate. The two extreme strategies may well represent adaptive peaks.

LITERATURE CITED

ALEXANDER, R. D. 1975. Natural selection and specialized chorusing behavior in acoustical insects. *In*: Pimentel, D. (ed.). Insects, Science, and Society. Academic Press, New York. Pp. 35–77.

ARP, W. 1957. Observaciones sobre el comportamiento en grupo del *Phaethornis longuemareus* (Aves-Trochilidae). Bol. Soc. Venezolana Hist. Nat. 17: 156–168.

BROWN, J. L. 1964. The evolution of diversity in avian territorial systems. Wilson Bull. 76: 160–169.

BUECHNER, H. K. 1961. Territorial behavior in Uganda kob. Science 133: 698–699.

———, AND H. D. ROTH. 1974. The lek system in Uganda kob antelope. Amer. Zool. 14: 145–162.

CAMPANELLA, P. J., AND L. L. WOLF. 1974. Temporal leks as a mating system in a temperate zone dragonfly (Odonata: Anisoptera). I. *Plathemis lydia* (Drury). Behaviour 51: 49–87.

CODY, M. L. 1968. Interspecific territoriality among hummingbird species. Condor 70: 270–271.

DANIELS, G. S., AND F. G. STILES. 1979. The *Heliconia* taxa of Costa Rica: keys and descriptions. Brenesia 15, suppl. 1: 1–150.

DAVIS, T. A. W. 1934. Notes on display in the hummingbirds *Phaethornis superciliosus* (Linn.) and *Pygmornis ruber* (Linn.). Ibis 4: 732–738.

———. 1958. The displays and nests of three forest hummingbirds of British Guiana. Ibis 100: 31–39.

DUELLMAN, W. E. 1967. Social organization in the mating calls of some neotropical anurans. Amer. Midl. Naturalist 77: 156–163.

EMLEN, S. T. 1976. Lek organization and mating strategies in the bullfrog. Behav. Ecol. Sociobiol. 1: 283–313.

FFRENCH, R. 1973. A guide to the birds of Trinidad and Tobago. Wynnewood, Pennsylvania, Livingston Publ. Co.

FOGDEN, M. P. L. 1972. The seasonality and population dynamics of equatorial forest birds in Sarawak. Ibis 114: 307–343.

GILL, F. B., AND L. L. WOLF. 1975. Economics of feeding territoriality in the golden-winged sunbird. Ecology 56: 333–345.

GILLIARD, E. T. 1962. On the breeding behavior of the Cock-of-the-Rock (Aves, *Rupicola rupicola*). Bull. Amer. Mus. Nat. Hist. 124: 31–68.

HAINSWORTH, F. R., AND L. L. WOLF. 1972a. Energetics of nectar extraction in a small, high altitude, tropical hummingbird, *Selasphorus flammula*. J. Comp. Physiol. 80: 377–387.

———, AND ———. 1972b. Crop volume, nectar concentration and hummingbird energetics. Comp. Biochem. Physiol. 42A: 359–366.

HARTZLER, J. E. 1974. Predation and the daily timing of Sage Grouse leks. Auk 91: 532–536.

HOGAN-WARBURG, A. J. 1966. Social behavior of the Ruff, *Philomachus pugnax* (L.). Ardea 54: 109–229.

HOLDRIDGE, L. R., W. C. GRENKE, W. H. HATHEWAY, T. LIANG, AND J. A. TOSI, JR. 1971. Forest environments in tropical life zones: A pilot study. New York, Pergamon Press.

JANZEN, D. H. 1971. Euglossine bees as long-distance pollinators of tropical plants. Science 171: 203–205.

JEHL, J. R., JR. 1970. Sexual selection for size differences in two species of sandpipers. Evolution 24: 311–319.

KRUIJT, J. P., AND J. A. HOGAN. 1967. Social behavior on the lek in the black grouse, *Lyrurus tetrix tetrix* (L.). Ardea 55: 203–240.

———, C. J. DEVOS, AND I. BOSSEMA. 1972. The arena system of Black Grouse, *Lyrurus tetrix tetrix* (L.). Proc. 15th Intern. Ornithol. Congr.: 399–423.

LACK, D. 1968. Ecological adaptations for breeding in birds. London, Methuen and Co., Ltd.

LEUTHOLD, W. 1966. Variation in territorial behavior of Uganda kob *Adenota kob thomasi* (Neumann 1869). Behaviour 27: 214–257.

LILL, A. 1966. Some observations on social organization and nonrandom mating in captive Burmese red junglefowl (*Gallus gallus spadiceus*). Behaviour 26: 228–242.

———. 1974. Sexual behavior of the lek-forming White-bearded Manakin (*Manacus manacus trinitatis* Hartert). Z. Tierpsychol. 36: 1–36.

———. 1976. Lek behavior in the golden-headed manakin, Pipra erythrocephala in Trinidad (West Indies). Adv. in Ethology, Suppl. 18 to J. Comp. Ethology.

LINHART, Y. B. 1973. Ecological and behavioral determinants of pollen dispersal in hummingbird-pollinated *Heliconia*. Amer. Naturalist 107: 511–523.

MEYER DE SCHAUENSEE, R. 1966. The species of birds of South America and their distributions. Narberth, Pennsylvania, Livingston Press.

NICHOLSON, E. M. 1931. Communal display in hummingbirds. Ibis 1: 73–83.

ONIKI, Y. 1970. Nesting behavior of reddish hermits (*Phaethornis ruber*) and occurrence of wasp cells in nests. Auk 87: 720–728.

ORIANS, G. H. 1969. On the evolution of mating systems in birds and mammals. Amer. Naturalist 105: 589–603.

ORTIZ-CRESPO, F. I. 1972. A new method to separate adult and immature hummingbirds. Auk 89: 851–857.

PARKER, G. A. 1970. The reproductive behavior and the nature of sexual selection in *Scatophaga stercoraria* L. (Diptera: Scatophagidae). I and II. J. Anim. Ecol. 39: 185–228.

PITELKA, F. A. 1942. Territoriality and related problems in North American hummingbirds. Condor 44: 189–204.

———, R. T. HOLMES, AND S. F. MACLEAN, JR. 1974. Ecology and evolution of social organization in Arctic sandpipers. Amer. Zool. 14: 185–204.

ROBEL, R. J. 1966. Booming territory size and mating success of the Greater Prairie Chicken (*Tympanuchus cupido pinnatus*). Anim. Behav. 14: 328–331.

———. 1967. Significance of booming grounds of Greater Prairie Chickens. Proc. Amer. Phil. Soc. 111: 109–114.

RUSCHI, A. 1950. O território e as áreas de alimentação e de nidificação de *Anisopterus* pretrei (DeLattre and Lesson), observadas através algumas gerações. Bol. Mus. Biol. Prof. Mello-Leitão 8: 1–20.

———. 1961. Algumas observacoes sobre: *Phaethornis yaruqui yaruqui* (Bourcier); etc. Bol. Mus. Biol. Prof. Mello-Leitão 27: 1–21.

———. 1967. A plumagem e a muda em *Phaethornis idaliae*. Bol. Mus. Biol. Prof. Mello-Leitão 33: 1–5.

SCHOENER, T. W. 1971. Theory of feeding strategies. Ann. Rev. Ecol. Syst. 2: 369–404.

SIBLEY, C. G. 1957. The evolution and taxonomic significance of sexual dimorphism and hybridization in birds. Condor 59: 166–191.

SKUTCH, A. F. 1951. Life history of Longuemare's Hermit Hummingbird. Ibis 93: 180–195.

———. 1958. Life history of the Violet-headed Hummingbird. Wilson Bull. 70: 5–19.

———. 1964a. Life histories of hermit hummingbirds. Auk 81: 5–25.

———. 1964b. Life history of the Scaly-breasted Hummingbird. Condor 66: 186–198.

———. 1966. A breeding bird census and nesting success in Central America. Ibis 108: 1–16.

———. 1967. Life histories of Central American highland birds. Publ. Nuttall Ornithol. Club No. 7.

SLUD, P. 1960. The birds of Finca "La Selva," a tropical wet forest locality. Bull. Amer. Mus. Nat. Hist. 121: 49–148.

———. 1964. The birds of Costa Rica: distribution and ecology. Bull. Amer. Mus. Nat. Hist. 128: 1–430.

SMITH, S. M. 1976. A field study of dominance hierarchies in Black-capped Chickadees. Auk 93: 95–107.

SNOW, B. K. 1972. A field study of the calfbird *Perissocephalus tricolor*. Ibis 114: 139–162.

———. 1973a. Social organization of the hairy hermit *Glaucis hirsuta*. Ardea 61: 94–105.

———. 1973b. The behavior and ecology of hermit hummingbirds in the Kanaku Mountains, Guyana. Wilson Bull. 85: 163–177.

———. 1974. Lek behaviour and breeding of Guy's Hermit Hummingbird *Phaethornis guy*. Ibis 116: 278–297.

———, AND D. W. SNOW. 1972. Feeding niches of hummingbirds in a Trinidad valley. J. Anim. Ecol. 41: 471–485.

SNOW, D. W. 1962. A field study of the Black and White Manakin, *Manacus manacus*, in Trinidad. Zoologica 47: 65–104.

———. 1963. The evolution of manakin displays. Proc. 13th Intern. Ornithol. Congr.: 553–561.

———. 1968. The singing assemblies of little hermits. Living Bird 7: 47–55.

———, AND B. K. SNOW. 1964. Breeding seasons and annual cycles of Trinidad land-birds. Zoologica 49: 1–39.

———, AND ———. 1973. The breeding of the Hairy Hermit *Glaucis hirsuta* in Trinidad. Ardea 61: 106–122.

STILES, F. G. 1971a. Time, energy, and territoriality of the Anna Hummingbird (*Calypte anna*). Science 171: 818–821.

———. 1971b. On the field identification of California hummingbirds. California Birds 2: 41–54.

———. 1973. Food supply and the annual cycle of the Anna Hummingbird. Univ. California Publ. Zool. 97: 1–109.

———. 1975. Ecology, flowering phenology, and hummingbird pollination of some Costa Rican *Heliconia* species. Ecology 56: 285–302.

———. 1978a. Temporal organization of flowering among the hummingbird food plants of a tropical wet forest. Biotropica 10: 194–210.

———. 1978b. Possible specialization for hummingbird-hunting in the Tiny Hawk. Auk 95: 550–553.

———, AND L. L. WOLF. 1970. Hummingbird territoriality at a tropical flowering tree. Auk 87: 467–491.

———, AND ———. 1973. Techniques for color-marking hummingbirds. Condor 75: 244–245.

———, AND ———. 1974. A possible circannual molt rhythm in a tropical hummingbird. Amer. Naturalist 108: 341–354.

VAN RHIJN, J. 1973. Behavioral dimorphism in male Ruffs, *Philomachus pugnax*. Behaviour 47: 153–229.

VAN TYNE, J., AND A. J. BERGER. 1959. Fundamentals of ornithology. New York, John Wiley and Sons, Inc.

WARD, P. 1969. The annual cycle of the Yellow-vented Bulbul *Pycnonotus goiavier* in a humid equatorial environment. J. Zool. 157: 25–45.

WASER, N. M., AND CALDER, W. A. 1975. Possible impairment of nest-building of hummingbirds by acetate leg tags. Condor 77: 361.

WILEY, R. H. 1971. Song groups in a singing assembly of Little Hermits. Condor 73: 28–35.

———. 1973. Territorially and non-random mating in Sage Grouse, *Centrocercus urophasianus*. Anim. Behav. Monogr. 6: 87–169.

———. 1974. Evolution of social organization and life-history patterns among grouse. Quart. Rev. Biol. 49: 201–227.

WOLF, L. L. 1969. Female territoriality in a tropical hummingbird. Auk 86: 490–504.

———. 1970. The influence of seasonal flowering on the biology of some tropical hummingbirds. Condor 72: 1–14.

———. 1975a. "Prostitution" behavior in a tropical hummingbird. Condor 77: 140–144.

———. 1975b. Female territoriality in the Purple-throated Carib. Auk 92: 511–522.

———, F. R. HAINSWORTH, AND F. B. GILL. 1975. Foraging efficiencies and time budgets in nectar feeding birds. Ecology 56: 117–128.

———, ———, AND F. G. STILES. 1972. Energetics of foraging: rate and efficiency of nectar extraction by hummingbirds. Science 176: 1351–1352.

———, AND F. G. STILES. 1970. Evolution of pair cooperation in a tropical hummingbird. Evolution 24: 759–774.

———, ———, AND F. R. HAINSWORTH. 1976. Ecological organization of a highland, tropical hummingbird community. J. Anim. Ecol. 45: 349–379.

———, AND J. S. WOLF. 1976. Mating system and reproductive strategy of malachite sunbirds. Condor 78: 27–39.

WYNNE-EDWARDS, V. C. 1962. Animal dispersion in relation to social behaviour. Edinburgh, Oliver and Boyd.

Heliconia species designated by taxon number in Tables 4–7, Figure 20, and elsewhere in the text are described in a recent publication (Daniels and Stiles 1979) that appeared too late for the names to be changed here. Names assigned are: H-3 = *H. mathiasii*; H-16 = *H. umbrophila;* H-17 = *H. sarapiquensis;* H-18 = *H. irrasa. H. sublata* of Figure 9 = *H. mathiasii.*—Auth.